Environmental Issues in Political Discourse in Britain and Ireland

Environmental Issues in Political Discourse in Britain and Ireland

Edited by

Gilles Leydier and Alexia Martin

CAMBRIDGE
SCHOLARS
PUBLISHING

Environmental Issues in Political Discourse in Britain and Ireland,
Edited by Gilles Leydier and Alexia Martin

This book first published 2013

Cambridge Scholars Publishing

12 Back Chapman Street, Newcastle upon Tyne, NE6 2XX, UK

British Library Cataloguing in Publication Data
A catalogue record for this book is available from the British Library

ISBN (10): 1-4438-4851-4, ISBN (13): 978-1-4438-4851-0

TABLE OF CONTENTS

INTRODUCTION

Environmental issues have gained increasing prominence in public affairs in modern societies since the end of the 1960s, leading to the establishment and development of various social movements, NGOs and political parties. The ecology movement became politicized throughout the British Isles in the 1970s, following the rise of new social pressure groups (such as the Women's Movement and various pacifist movements). Yet, despite the increasing expression of concern in political and media debates about issues such as climate change, pollution and threats to biodiversity, "political ecology" (operating at the confluence of scientific developments, political engagement and ethical debates) is still trying to find its bearings.

In Britain, while the Green party has remained conspicuously weak in terms of its electoral results, environmental issues have been progressively assimilated into the speeches and manifestos of the main political parties. Spurred by the example set by New Labour, the Tories and the Liberal Democrats "turned green", vying to show initiative in measures to reduce greenhouse gas emissions (following the Kyoto agreement) in formal manifesto pledges or press releases.

Moreover, questions relating to sustainable development and the management of natural resources now play a key role in the relations between Westminster and various organizations and authorities in Scotland, Wales and Ireland – where local attachment to landscape is a unifying factor and nationalist parties have tended to link issues relating to environmental protection with questions of identity in their actions on both national and European stages. Thus, since the 1990s, nationalists and ecologists have come together and signed election alliances, particularly in relation to the nuclear issue.

In Ireland, the ecologists' winning streak during the 2007 general elections was short-lived as their time in office was curtailed by the worst crisis the country has known since gaining its independence. Despite their resignation from government a few months before the general elections of 2011, the Green Party lost half their voters and their six MPs at the polls.

This volume seeks to examine the political dimension of environmental concerns in the context of the British Isles, notably how these issues have been assimilated by political parties, which issues have been given priority, the scope and nature of the influence of the main actors, and the role of ecologists. It is a collective work bringing together researchers and academics from a range of disciplines and backgrounds, who offer diverse perspectives on the subject. They aim at analyzing the political response to current environmental issues, revealing some cases of genuine commitment and effective action, and sometimes highlighting discrepancies between statements of intent and election pledges on the one hand, and implemented policies on the other.

The book is divided into six main sections. The first part of the volume clarifies some of the theories and concepts found in contemporary political ecology. To start with, Clare Sibley-Esposito provides an in-depth examination of the concept of 'natural capital' – defined in the field of ecological economics as the stock of physical natural assets which provide a range of 'ecosystem services', ranging from 'carbon sequestration' to the 'pollination services' provided by bees. The author shows how this concept has been taken up in recent years in Britain, as part of an international agenda to develop more extensive economic valuation of natural habitats and processes. She starts by showing how the natural environment has come to be seen by certain policy-makers as a 'service provider', interconnected to economic systems, following calls by ecological economists for greater attention to be paid to the previously invisible economic values of such environmental factors, as a means of encouraging conservation measures. The concept has been wielded in a number of ways, including as part of attempts to emphasize the shortcomings of GDP as an indicator of national wealth, with plans having been drawn up to work towards taking 'natural capital' into account on national balance sheets.

This fully-documented first chapter not only provides useful insights to grasp this concept, but also refers to a number of key British and international studies and policies which testify to widespread statements of intention to adopt 'natural capital accounting' on the part of British authorities, both in Westminster and the Devolved Administrations, and also in the private sector. Introduced by the previous government, the idea was highlighted by David Miliband in 2006, then by the coalition government led by David Cameron, in statements clearly acknowledging that economic activity is dependent on natural assets. Despite underlining this relatively broad political consensus, the author does not fail to look into the ideological and methodological issues raised by such monetary valuations of Nature, which Green MP Caroline Lucas assimilates to man's exploitation of Nature.

In the second chapter ("Gaia: Myth beyond Science, Ethics beyond Politics"), Coralie Raffenne delves into a more controversial concept which illustrates the ethical dimension of the environmental discourse: the Gaia hypothesis. Associated with the charismatic and provocative chemist James Lovelock and with the deep ecology movement, the originality of this hypothesis rests on its multi-faceted approach, which is scientific, holistic and ethical. While drawing on ancient references and myths (Greek goddess Gaia, Mother Earth…), it is also a scientifically-demonstrated vision of the Earth as a self-regulated system, able to keep its own equilibrium and adapt to change, whatever perturbations might be generated by human interventions.

In the context of political ecology, the author shows how Lovelock's proposition calls for us to reconsider our relationship with the environment, not only in economic terms but also ethically-speaking. Coralie Raffenne also notes the potential for political action derived from Lovelock's theories, with propositions "to put democracy on hold for a while" in states of 'green' emergency (climate change, threats to biodiversity…) or to take global and radical measures. The unclassifiable scientist opposes mainstream environmentalist approaches, which support local and participatory decision-making.

The second part of the volume focuses on discourses on and representations of the environmental issue in the public sphere. To start with Camille Biros questions the identity and legitimacy of these new experts in the communication of organizations in the United Kingdom from the beginning of the years 2000 to the present. After having given a definition of the notion of expertise in the rather young environment field, she demonstrates that the usual criteria used for identifying experts have a limited efficiency in this case, and then explores discourse analysis tools in order to draft characteristics of the environment expert. Her contribution, based on the corpus of Corporate Social Responsibility reports, analyses the status and work of some key players who can be considered as prototypes of the expert in the environment field.

From a different perspective, Pierre-François Peirano addresses the environment debate in post-industrial countries through its representation in the popular television series *Yes, Minister*, broadcast by the BBC in the early 1980s. If the issue of the environment in the British series provides comic relief and is used as a pretext to ridicule politicians, the argument is that its representation is less naïve than it seems and calls for awareness on the part of politicians. And the overwhelming role of the environment in the series accounts for its newly-acquired importance on the political scene.

The third part of the volume concentrates on the present political debate in the United Kingdom. In his contribution, Neil Carter explores the party politicization of climate and energy policy in Britain over the last decade. After having discussed the environment legacy of the successive New Labour governments, he analyses the dramatic rise of the climate change issue in the domestic political agenda after 2006. He then questions whether the three-party competition around environment issues has contributed to the radicalisation of climate policy over the last years. Eventually, he discusses the sustainability of the new political consensual for progressive climate policy under the present Coalition government.

Hélène Ledouble and Olivier Gouirand for their part focus on Prime Ministerial discourse and the way the environment question is tackled. Their corpus includes speeches delivered by Blair, Brown and Cameron during their time in office. Using discourse analysis methodology, they uncover the disparity in the lexical as well as semantic richness of the environmental terms at hand. Their diachronic approach demonstrates the changes in the use of some prominent lexical units of environmental discourse or even the loss thereof over the last years. The reasons that account for these changes are also addressed, and the argumentative stance of the Prime Ministerial speeches is eventually decoded.

From another perspective, Muriel Cassel-Piccot analyses how the British Liberal Democrats have taken up the green cause since the 1970s. By exploring the party's policy proposals and manifesto pledges as well as the evolution of the activists and voters' perception of the environment issue, she assesses the greening of the yellow party from various ways and analyzes to what extent the greening process has been a source of internal cohesion and/or division. She argues that although the party's environmental credentials are undisputable and its reputation for being the greenest of the three main British parties justified, its greening has been relative. The Liberal Democrats have given greater priority to the adjustment of their green shade to the current global economy than to an unprecedented deepening of the hue.

The fourth part of the volume moves to the 'Celtic periphery' of Great Britain in order to explore the environmental issue in the Scottish and Welsh contexts. In her contribution, Stephanie Bory reminds that environmental issues have represented a nationalistic niche in Wales for several decades since questions of identity and the protection of the environment and landscape have been deeply interconnected. And indeed after twelve years of existence the Welsh institutions have had recourse to environmental issues to assert and acknowledge Wales'specificity, or even her superiority, both inside the UK and in Europe. However faced with institutional as well as economic obstacles the National Assembly for Wales has partly failed to turn words into concrete actions and Welsh people are more and more turning to nongovernmental organisations in order to deliver the green speech.

As far as Scotland is concerned, Fiona Simpkins argues that energy and environment issues have been a contentious problem for several years in the debate opposing London to the Scottish nationalist government. In adopting diverging strategies the SNP government aim at withdrawing Scotland from Westminster's sphere of influence and promote a unique Scottish environmental model refusing nuclear energy and championing renewable energy sources as well as a low carbon economy. Her contribution explores in detail the potential and the credibility of this alternative model for a 'Greener Scotland', and analyses the extent to which environmental concerns serve the nationalists to fuel the Scottish separatist debate.

The fifth part takes us to the other side of the Irish Sea to explore the political response to environmental issues in the context of the most western of the British Isles. In "Ireland after the Boom and Bust", Frank McDonald deplores some of the ill effects of the "Celtic Tiger" on the Republic of Ireland's countryside and urban landscape. Fuelled by lucrative tax incentives, construction output reached unprecedented levels in the early 2000s while prices rose to staggering highs. This chapter starts with an uncompromising examination of the environmental impact of such a housing boom, from the mushrooming of "ghost housing estates" (when the bubble burst in 2008) and ostentatious "McMansions" in rural areas, to the construction of luxury hotels and new roads near historical sites.

The author of *Chaos at the Crossroads* (Gandon editions, 2005) also draws attention to some of the political decisions which have fostered this situation and to the role played by the financial system. To illustrate his point, he shows the failings and intricacies of Dublin's urban planning and bureaucracy, decrying an "incredibly weak" civic governance that led to the approval of several schemes for high-rise and disproportionate buildings, quite foreign to Dublin's skyline and character – most of which were abandoned after the property crash.

Another highly sensitive issue of the last decades in Ireland is certainly the debate about nuclear energy and the controversy over the British nuclear site of Sellafield. Through the study of the Irish political discourse and policies on nuclear energy since the 1970s, Alexia Martin sheds light on the various aspects of the Irish antinuclearism, from a global, European and Anglo-Irish perspective. She shows how, after abandoning the attempt to introduce nuclear power in Ireland, successive Irish governments eventually rallied behind public opinion to oppose the potential impact of the nuclear reprocessing activities in Sellafield. The diplomatic and legal battle of the 1990s and early 2000s leads the author to examine the deeper identity-based motives of the controversy, in the face of the legacy of colonialism and European homogenization.

Questions relating to Irish identity are also at the heart of Frédéric Armao's research paper, entitled "The Color Green in Ireland: Ecological Mythology and the Recycling of Identity". The author deploys the concept of "ecological mythology" to consider how, beyond the usual association of the Emerald Isle with the colour green, symbolic meanings have been recycled through the ages to suit first religious, then political and marketing ends. This thorough study offers a fresh perspective on aspects of the "Irish green" which we tend to take for granted, from Ireland's lush vegetation that has inspired poets throughout the ages to Saint Patrick, shamrocks and the republican struggle against loyalism. The author not only reveals the historical background of this potent visual representation of the Republic of Ireland (its 'unofficial color'), but also takes stock of the numerous instances and hijacking of this symbol in contemporary political discourse.

Lastly, the sixth part of the volume deals with the salient issue of climate change from two different yet complementary angles. Nicholas Sowels presents a meticulous examination of the Stern Review, a major study of the economic impact of climate change, commissioned by Gordon Brown (then Chancellor of the Exchequer) and published in 2006. He starts by a detailed study of the findings of the Review, its methodological approach and its key recommendations, which highlight the urgent need for intervention both by the UK government and the international community as a whole. He goes on to examine how the Stern Review was received in political and economic spheres – ranging from praise to criticism – and how it was concretely addressed by political leaders, from New Labour to the Conservative-Liberal Democrat coalition. The author demonstrates how the Review "provided additional impetus" to existing policies against climate change, and contributed to placing Britain at the forefront of international moves to set emission reduction targets.

If the Stern Review works on the assumption of the reality of climate change, some scientists have challenged the very legitimacy of the debate or have voiced their disagreement concerning the grounds for identifying a link between global warming and human activities. Scientific controversies provide the starting point for Patrick Menneteau's reflections concerning the issue, ("Global Warming : between Doubts and Certainty") This original in-depth study approaches this crucial debate from a new perspective derived from phenomenology and psychology, showing how subjective experience lies at the root of present-day scientific debate on climate change, beyond scientific and ideological motives. The methodological and rhetorical tools used by the opposing scientific camps are rigorously assessed, in the light of philosophy and psychology, so as to uncover the modes of conceptualization and the archetypes influencing various types of discourse, leading the author to call for a change of perspective on the issue.

PART I:

POLITICAL ECOLOGY: THEORIES AND CONCEPTS

ACCOUNTING FOR ECOLOGY? THE CONCEPT OF 'NATURAL CAPITAL' IN BRITISH POLITICAL DISCOURSE AND POLICY, 1997-2012

CLARE SIBLEY-ESPOSITO

UNIVERSITY OF TOULON, BABEL RESEARCH LABORATORY

> "…the modern industrial system, with all its intellectual sophistication, consumes the very basis on which it has been erected. To use the language of the economist, it lives on irreplaceable capital which it cheerfully treats as income."
>
> E.F Schumacher, *Small is Beautiful: A study of economics as if people mattered,* 1973[1].

> "Natural capital can be defined as the stock of our physical natural assets (such as soil, forests, water and biodiversity) which provide flows of services that benefit people (such as pollinating crops, natural hazard protection, climate regulation or the mental health benefits of a walk in the park). Natural capital is valuable to our economy. Some marketable products such as timber have a financial value that has been known for centuries. In other cases (e.g. the role of bees in pollinating crops), we are only just beginning to understand their financial value."
>
> English Natural Environment White Paper, *The Natural Choice: Securing the Value of Nature*, 2011[2].

First wielded in the early 1970s by E.F Schumacher in his famous call to reflection on the potential unsustainability of modern Western modes of consumption, by 2012 the concept of 'natural capital' had emerged as a cornerstone of a number of government research studies and policies being developed around the United Kingdom, partly in response to several

---

[1] Schumacher, E.F, *Small is Beautiful: A study of economics as if people mattered*, London, Blond and Briggs, 1973: 17.

[2] English Natural Environment White Paper, *The Natural Choice: Securing the Value of Nature*, June 2011: 11. http://www.official-documents.gov.uk/document/cm80/8082/8082.pdf

international projects. Yet, whereas for Schumacher the representation of the natural world as an asset-yielding stock served as a metaphor which involved *borrowing* "the language of the economist", early twenty-first century Britain has seen attempts to develop the concept of 'natural capital' in such a way as to *integrate it into* the current socio-economic paradigm, by enabling financial values to be assigned not only to raw materials but also to a range of 'ecosystem services' - from the 'carbon sequestration' and 'flood prevention services' offered by certain types of fluvial wetland to the 'pollination services' provided by bees. In a White Paper focusing on environmental issues, published in June 2011, the Coalition government led by David Cameron announced a "commitment to putting natural capital at the centre of economic thinking"[3], along with its intention to set up a Natural Capital Committee and an Ecosystems Markets Task Force. The Welsh Assembly government issued a statement in support of this approach, emphasising that 'natural capital accounting' is evoked in the Welsh Natural Environment Framework, set up in 2010[4]. Meanwhile, Scottish Natural Heritage continued to develop its Natural Capital Asset (NCA) index, launched in March 2011 with the assertion that Scotland was "the first country in the world to publish such a detailed attempt to measure annual changes in its natural capital based on an evaluation of ecosystem services"[5].

This chapter focuses on tracing the main stages in the process by which the concept of 'natural capital' came to be mainstreamed in policy circles in the UK by 2012, as part of an international agenda. It goes on to consider the dilemma which the concept may pose to environmentalists, briefly evoking some of the methodological issues raised by attempts to develop the means by which to put a range of habitats and natural processes on both national and corporate balance sheets.

The birth of a concept: the natural environment as 'service provider'

In May 1997, just as Tony Blair was settling into Downing Street following Labour's first general election victory in over twenty years, the

---

[3] *Ibid.*

[4] Welsh Assembly Government Written Statement: *The Natural Environment Framework 'A Living Wales'*, 15 June 2011.
http://wales.gov.uk/about/cabinet/cabinetstatements/2011/110615nat/?lang=en

[5] Scottish Natural Heritage, *Scotland's Natural Capital Asset (NCA) Index, 2012 version*: 1.
http://www.snh.gov.uk/docs/B814140.pdf

renowned journal *Nature* published a landmark article which was to generate debate in the scientific community for years to come. Entitled 'The value of the world's ecosystem services and natural capital', this controversial paper literally put a price on the economic benefits accrued from global natural resource stocks and ecosystem processes, estimating at the time such 'natural capital' and 'ecosystem services' to have an annual economic value of $33 trillion, nearly twice the total sum of the GDPs of the world's nations[6]. Whilst acknowledging the "conceptual and empirical problems inherent in producing such an estimate", the authors stressed that they were attempting to provide a synthesis of work undertaken in the emerging field of ecological economics, in order to draw attention to the extent of the dependence of economic systems on environmental factors[7]. By attempting to calculate the "hidden values" associated with seventeen categories of "ecosystem services", ranging from the regulation of atmospheric gases to "cultural values", they sought to stimulate more widespread economic assessment of such factors, arguing that:

> Because ecosystem services are not fully 'captured' in commercial markets or adequately quantified in terms comparable with economic services and manufactured capital, they are often given too little weight in policy decisions.[8]

This argument was debated in scientific circles for several years before gradually being taken up by policy-makers[9]. The publication in 2005 of the findings of the Millennium Ecosystems Assessment (MA), a four-year international study led by the United Nations Environment Programme, is frequently cited as marking a turning point, following which the concept of 'natural capital accountancy' began to gain international political currency. The MA found "nearly two thirds of the services provided by nature to humankind [...] to be in decline world-wide", asserting that "the

---

[6] Costanza, R, d'Arge, R, de Groot,R, Farber,S, Grasso,M, Hannon,B, Limburg,K, Naeem,S, O'Neill,RV, Paruelo,J, Raskin,RG, Sutton,P, Van den Belt,M: 'The value of the world's ecosystem services and natural capital.' *Nature*, 1997, 387, 253-60. http://www.nature.com/nature/journal/v387/n6630/full/387253a0.html

[7] *Ibid.*, 253.

For a detailed survey of the development of the field of ecological economics, see: Goméz-Baggethun, *et al.*, 'The history of ecosystem services in economic theory and practice: from early notions to markets and payment schemes', *Ecological Economics*, 2009. doi:10.1016/j.ecolecon.2009.11.007

[8] Costanza *et al.*: 253.

[9] Goméz-Baggethun, *et al.*, 'The history of ecosystem services in economic theory and practice'.

benefits reaped from our engineering of the planet have been achieved by running down natural capital assets"[10]. The study considered the development of techniques for the monetary valuation of ecosystem stocks and processes to be a means to "correct the historical bias that has existed against natural services when it comes to weighing the costs and benefits of particular economic choices - whether for individuals, businesses, or governments."[11] It also contributed to emerging debates about the shortcomings of GDP as an indicator of national wealth, by asserting that in a significant number of cases countries considered in conventional terms to be experiencing economic growth were actually becoming poorer when estimates of the economic costs of loss of their 'natural capital' were added to their balance sheets[12].

A perusal of parliamentary and media archives available on the Internet shows up expressions of concern about the depreciation of 'natural capital' going back to the late 1990s in the UK[13], whilst references to the MA occur in numerous documents produced by Defra and its advisory bodies from 2006 onwards[14]. However, the possibility of attributing monetary values to a wide range of environmental factors only began to make widespread media headlines following the publication of the UK's National Ecosystems Assessment (UK NEA), which included qualitative and quantitative evaluations of a range of 'ecosystem services'[15]. Originally commissioned by Gordon Brown's Labour government in 2009, the UK NEA was presented by the Coalition government in June 2011 as marking "a vital step forward in our ability to understand the true

---

[10] The Millenium Ecosystem Assessment Board, *Living beyond our means: natural assets and human well-being. Statement from the Board*, 2005: 5.
http://www.millenniumassessment.org/documents/document.429.aspx.pdf

[11] *Ibid.*, 22.

[12] *Ibid.*

[13] See, for example, the Second Report of the House of Commons Environmental Audit Committee for the 1997-1998 sessions, which drew on work by Friends of the Earth in calling for a revision of indicators to measure environmental damage and "the depreciation of natural capital".
http://www.publications.parliament.uk/pa/cm199798/cmselect/cmenvaud/517/51702.htm

[14] See, for example, 'An Introductory Guide to Valuing Ecosystem Services' published by the Department for Environment, Food and Rural Affairs in 2007 (Product code: PB12852). http://archive.defra.gov.uk/environment/policy/natural-environ/documents/eco-valuing.pdf

[15] *UK National Ecosystem Assessment: Understanding nature's value to society. Synthesis of key findings*. Cambridge, UNEP-WCMC, 2011.
http://uknea.unep-wcmc.org/Home/tabid/38/Default.aspx

value of nature and the how to sustain the benefits it gives us."[16] In the intervening period, a small number of MPs from all three main political parties were involved in attempting to give greater prominence to the concept of 'natural capital', with Barry Gardiner (Lab), Dr Alan Whitehead (Lab), Elliot Morley (Lab), Richard Benyon (Con), Zac Goldsmith (Con) and Dr Evan Harris (LibDems) among those who evoked it in parliamentary debates[17].

Some of the most high profile early attempts to draw attention to the concept were made by David Miliband during his stint as Secretary of State for Environment, Food and Rural Affairs, under Tony Blair's leadership[18]. In a speech to the Fabian Society in December 2006, Miliband suggested that integrating the concept of 'natural capital' into party policy constituted a major new challenge for Labour, ten years on from their return to government:

> [...]in 1997 we made economic stability and high employment our top priorities, but in 2007, we need a third ambition, to redress the imbalance between the natural resources we consume, and the natural capital we reinvest. A kind of 'golden rule' to ensure we do not mortgage the futures of our children in an unsustainable ecological debt. [...] In 1997, we said we wanted to extend the power of choice and voice that exist in the private sector to public services. Today, we need to extend market mechanisms to public goods. We need to put a price on carbon dioxide and use the power of the market to find the lowest cost emissions.[19]

Miliband drew on the *Stern Review* to bolster his argument in favour of carbon-trading. Published two months earlier, the Review had brought

---

[16] Caroline Spelman, Secretary of State for Environment, Food and Rural Affairs from 12 May 2010 to 4 September 2012, quoted in a report by Sarah Morrison in *The Independent*, 'What Price Nature? Report puts financial value on UK's ecology', 2 June 2011. http://www.independent.co.uk/environment/nature/what-price-nature-report-puts-financial-value-on-uks-ecology-2292043.html

[17] See, for example, the transcript of the parliamentary debate which took place in Westminster Hall on 30 April 2009, concerning the government's involvement in the Darwin Initiative.
http://www.publications.parliament.uk/pa/cm200809/cmhansrd/cm090430/halltext/90430h0001.htm

[18] David Miliband was Secretary of State for Environment, Food and Rural Affairs from 5 May 2006 to 27 June 2007.

[19] David Milband, 'Red-green renewal: the future of new Labour' in *Social Europe,* Winter 2007, 135.
http://www.social-europe.eu/fileadmin/user_upload/Authors/Miliband_2-3.pdf

economic considerations to the fore in debates about environmental policy, by concluding that the "benefits of strong, early action on climate change outweigh the costs."[20] It had called for the establishment of carbon prices, to be applied in the context of taxation, trading or regulation[21]; Miliband envisaged a mix of such measures as being the way forward to building a "low-carbon" future, with "a new market at its heart: a market for carbon, with the vast majority of the economy covered by carbon trading."[22]

A few months later, speaking at a meeting of the Campaign for the Protection of Rural England, Miliband evoked a wider conception of 'ecosystem service' valuation:

> [...]we are now beginning to value environment assets that in the past we have thought of as a free good. Carbon is the most obvious example. As well as valuing carbon emissions from fossil fuels, we need to think how we value carbon sinks from forests and peats. But carbon is just one environmental public good. As the Millennium Ecosystems Assessment report sets out, there are range of ecosystem services that regulate the climate, protect us from floods, purify water, and provide aesthetic and recreational value. [23]

The concept of 'natural capital accountancy' had been gaining momentum behind the scenes in Defra[24], and clearly the Secretary of State had been paying close attention to UN-sponsored developments in the field of ecological economics in the lead up to the G8+5 environment ministers' meeting which was to take place in Potsdam the following week, from 15-17 March 2007.

The international agreement referred to as the 'Potsdam Initiative'[25]

---

[20] Nicholas Stern, *The Economics of Climate Change : The Stern Review*, Cambridge, Cambridge University Press, 2007: 191.
[21] *Ibid.*, 351-366.
[22] David Miliband, 'Red-green renewal: the future of new Labour': 136.
[23] David Miliband, in 'A Land Fit for the Future', a speech made to the Campaign for the Protection of Rural England, London, 9 March 2007; quoted from the transcript of the speech included in the annexes of the Forestry Commission England National Committee Paper (15/07) Director Report, March 2007. http://www.forestry.gov.uk/pdf/enc-mar07-paper-15.pdf/$file/enc-mar07-paper-15.pdf
[24] See 'An Introductory Guide to Valuing Ecosystem Services', Defra, 2007.
[25] The title given at the time of the meeting was 'The Potsdam Initiative: Biological Diversity 2010'. The G8+5 Environment and Energy Ministers' meeting took place in Potsdam, Germany, 15-17 March 2007. The G8+5 refers to the G8 countries (Canada, France, Germany, Italy, Japan, Russia, the United

included a commitment to "examine the concept and viability of payments for ecosystem services" and to "approach the financial sector to effectively integrate biodiversity into its decision making", backed up by the funding of a global study to "initiate the process of analysing the global economic benefit of biological diversity, the costs of the loss of biodiversity and the failure to take protective measures versus the costs of effective conservation".[26] Entitled 'The Economics of Ecosystems and Biodiversity Study' (TEEB), the resulting research programme published its Interim Report in 2008, before presenting its main reports at the tenth Convention on Biodiversity[27] conference in Nagoya, Japan, in October 2010. TEEB then entered the currently on-going 'Implementation Phase', through which it seeks to work more closely with individual countries and organisations as part of its efforts to mainstream procedures for carrying out economic assessments of ecosystems and biodiversity[28].

TEEB's conclusions in 2010 attributed the rapid loss of global biodiversity in recent years to "the invisibility of many of nature's services to the economy" resulting in "widespread neglect of natural capital"[29]. The study's authors recommended that valuation techniques be further developed to enable previously unacknowledged input from these 'ecosystem services' to be integrated into cost-benefit analyses in all aspects of socio-economic activity, calling for both governments and companies to include such data in national accounts and annual reports[30]. They asserted that the use of monetary valuations would not automatically cause such 'natural assets' to become tradable commodities[31], but that both private and public stakeholders should share a "common language"[32] in order to enable the development of a range of approaches – including reforms of taxation and subsidies - designed to "help recalibrate the faulty economic compass that has led us to decisions that are prejudicial to both current well-being and that of future generations."[33]

Kingdom and the United States) along with Brazil, China, India, Mexico and South Africa.

[26] http://www.g8.utoronto.ca/environment/env070317-potsdam.htm

[27] The tenth Conference of the Parties to the Convention on Biological Diversity (CBD COP 10) took place in Nagoya (Aichi Prefecture), Japan, 18-29 October 2010.

[28] http://www.teebweb.org/about/

[29] *Ibid.*

[30] *Ibid.*, 26-27.

[31] *Ibid.*, 12.

[32] *Ibid.*, 24.

[33] *Ibid.*, 3.

The simultaneous growth of interest in the concept of 'natural capital' in the private and public sectors

As part of its ambition to encourage widespread adoption of 'natural capital accounting', TEEB published separate reports aimed at influencing different levels of public policy in addition to a document destined specifically for the business community. The latter puts forward figures suggesting that by 2010 a number of companies around the globe were beginning to recognise that paying attention to environmental considerations could make economic sense, be it, for example, by avoiding costs associated with cleaning up pollution, by tapping into growing consumer interest in 'green goods' or by investing in emerging markets for some of the newly identified 'ecosystem services'. The report argues that:

> […] companies that understand and manage the risks presented by biodiversity loss and ecosystem decline, that establish operational models that are flexible and resilient to these pressures, and that move quickly to seize business opportunities, are more likely to thrive. Just as climate change has stimulated carbon markets and new business models, biodiversity and ecosystem services also offer opportunities for investors and entrepreneurs. However, there is a need to agree priorities and adopt an agenda for action – by business leaders, accountancy bodies, governments and other stakeholders – otherwise significant change is unlikely.[34]

In the UK, the emerging interest of parts of the private sector for 'natural capital accounting' can be attested by the setting up of a number of at least partly corporate-sponsored initiatives which make reference to TEEB in their documentation, following the publication of the study's Interim Report in 2008. For example, in a report presented to the House of Commons in the midst of a global financial crisis, The Aldersgate group[35] argued that "the natural capital assets that lay the foundations for our economy and society should not be off-balance sheet items similar to the risk exposures and subsequent heavy losses incurred in the banking sector

[34] *The Economics of Ecosystems and Biodiversity Report for Business: Executive Summary*, 2010: 13. http://www.teebweb.org

[35] The Aldersgate group describes itself as "a coalition of private, public and third sector organisations who believe that high environmental standards are essential for long term economic growth and international competitiveness". The group includes members as diverse as Friends of the Earth, British Telecom, Tesco, United Utilities and the Institution of Civil Engineers. http://www.aldersgategroup.org

during the 2008 credit crunch"[36]. It called for 'natural capital accounting' to be used in developing a clearer framework of environmental regulation which, it argued, "would produce significant cost savings and boost competitiveness"[37]. The Natural Capital Initiative, which describes itself as "a leading UK forum bringing together scientists, policy-makers, business, industry and others", was launched in 2009[38], whilst in September 2010 the University of Cambridge set up a "business-led" Natural Capital research programme, on the basis that "the growing pressures on ecosystem services and natural capital are expected to generate a range of future risks and opportunities for businesses."[39]

Meanwhile efforts were being redoubled on the international scene, often partly with UK government financing, to develop methodological tools for 'natural capital accounting' in both public and private sector contexts. In 2010, at the Convention on Biological Diversity conference in Nagoya, the World Bank announced the launch of the 'Wealth Accounting and Valuation for Ecosystem Services' (WAVES) project - an international partnership, partly funded by the UK, which aims to work with Central Banks and Ministries of Finance and Planning around the globe "to promote sustainable development by ensuring that the national accounts used to measure and plan for economic growth include the value of natural resources".[40] Meanwhile, the Parliamentarians Forum at the conference endorsed a Natural Capital Action Plan[41], outlining a set of recommendations for ways governments could aim "to mainstream ecosystem valuation into the economic growth model" and move beyond reliance on traditional GDP accounting as a measure of economic well-being.

---

[36] Report by The Aldersgate Group, *Green Foundations 2009*: 4. http://www.aldersgategroup.org.uk/asset/download/117/green_foundations_2009.pdf

[37] *Ibid.*

[38] The Natural Capital Initiative (NCI) is a partnership between the Society of Biology, the Centre for Ecology and Hydrology, the British Ecological Society and the James Hutton Institute. The NCI was launched in 2009 at a symposium entitled 'Valuing Our Life Support Systems', bringing together 250 representatives from business, academia, non-profit and public organisations. http://www.naturalcapitalinitiative.org.uk/about

[39] 'The Cambridge Natural Capital Programme: Understanding and managing the business risks and opportunities relating to ecosystems and natural capital.' Report produced by the University of Cambridge Programme for Sustainability Leadership, 2011: 3. http://www.cpsl.cam.ac.uk

[40] http://www.wavespartnership.org/

[41] http://www.globeinternational.org.

The shared focus of these initiatives on developing economic valuations for an ever-widening range of natural habitats and processes is consistent with the emphasis on 'natural capital accounting' promoted by the TEEB study and by elements of the revised Strategic Plan for Biodiversity adopted in Nagoya. Presented in the wake of the announcement that no nation had succeeded in meeting targets to halt biodiversity loss by 2010, in the first of its targets for the 2011-2020 period - known collectively as the Aichi targets - the Strategy evokes the "values" of biodiversity in somewhat general terms; the notion of economic assessment appears more clearly in the second target, which explicitly calls for measures of such "values" to be incorporated into planning strategies and national accounts:

> Target 1: By 2020, at the latest, people are aware of the values of biodiversity and the steps they can take to conserve and use it sustainably.
> Target 2: By 2020, at the latest, biodiversity values have been integrated into national and local development and poverty reduction strategies and planning processes and are being incorporated into national accounting, as appropriate, and reporting systems.[42]

The two years following the setting of the Aichi targets saw a proliferation of projects building on this United Nations-led impetus towards exploring possibilities for 'ecosystem service' valuation and 'natural capital accounting' on various scales around the world. The European Commission revised its biodiversity strategy in terms which demonstrated an unambiguous adherence to this agenda, announcing as part of an action plan published in May 2011 that:

> Member States, with the assistance of the Commission, will map and assess the state of ecosystems and their services in their national territory by 2014, assess the economic value of such services, and promote the integration of these values into accounting and reporting systems at EU and national level.[43]

The conference in Nagoya had also been the setting for the first steps in a process involving the private sector which was to lead to the

[42] United Nations Environment Programme, *Strategic Plan for Biodiversity 2011-20 and the Aichi Targets: Living in harmony with nature*. Report produced by the Secretariat of the Convention on Biological Diversity, 2010.
http://www.cbd.int/doc/strategic-plan/2011-2020/Aichi-Targets-EN.pdf

[43] *The E.U Biodiversity Strategy to 2020*, Luxembourg, Publications Office of the European Union, 2011: 17.

announcement at the Rio+20 Earth Summit[44] of a commitment by a range of financial institutions from around the globe to "work towards integrating natural capital considerations into [...] financial products and services"[45]. Whilst notoriously few ground-breaking international political agreements were reached in Rio, UK Deputy Prime Minister Nick Clegg sought to represent moves taken during the summit by both corporations and governments towards implementing 'natural capital accounting' as marking some of the positive aspects of an otherwise somewhat disappointing outcome:

> While the Rio declaration [in the final summit outcome document] was not all that we would have wanted, this is the first time that a multilateral document expressing such strong support for the green economy has been agreed. [...]Rio+20 recognised that we need to develop broader measures of progress to complement GDP in order to take account of the natural assets that will contribute to future prosperity—so-called GDP-plus. [...]at Rio national Governments recognised the importance of working alongside businesses. Thanks in no small part to the leadership of UK firms, Rio recognised the role of corporate sustainability reporting to their shareholders and to prospective investors—something that would have been inconceivable even a year ago.[46]

In the lead-up to the summit, possibly in anticipation of David Cameron's much-commented absence from its proceedings, Clegg had positioned himself as preparing to "[push] for greater global protections for our natural assets"[47], whilst the Secretary of State for Environment and Rural Affairs, Caroline Spelman, had asserted that the UK was "taking the lead" in work on natural capital accounting[48]. Both Clegg and Spelman were

---

[44] Generally referred to as RIO+20, being held on the 20th anniversary of the 1992 Earth Summit, the United Nations Conference on Sustainable Development was held from 20 – 22 June 2012 in Rio de Janeiro, Brazil.

[45] 'The Natural Capital Declaration: A commitment by the financial sector for Rio+20 and beyond', 2012.
http://www.naturalcapitaldeclaration.org.

[46] Nick Clegg's statement to the House of Commons following Rio+20, 26 Jun 2012 : Column 161.
http://www.publications.parliament.uk/pa/cm201213/cmhansrd/cm120626/debtext/120626-001.htm#12062679000001

[47] Nick Clegg in a speech entitled 'The myth - green versus growth' given at the KPMG headquarters in Canary Wharf, London, 11 April 2012.
http://www.libdems.org.uk/speeches.aspx

[48] Caroline Spelman in her address to businesses and charities at London's Guildhall on the UK aims for Rio+20, on 9 February 2012.

keen to draw attention to certain policy commitments presented in the English Natural Environment White Paper which had been published a year ahead of the Rio conference, in June 2011. These included the announcement of the intention to establish a Natural Capital Committee, to report to the Economic Affairs Committee (chaired by the Chancellor of the Exchequer), which would be involved in taking action "to capture the value of natural capital on the nation's balance sheet" as part of work towards developing measures of national wealth and well-being beyond traditional GDP indicators[49].

In seeking to draw attention to the concept of 'natural capital' as part of a stated ambition to "show environmental leadership internationally and within the E.U"[50], the Coalition government was brandishing a banner which had already been some time in the making, and which owed its existence in part to preparatory work carried out by the previous government and by various international organisations, as we have seen. However, the English Natural Environment White Paper is clear about its debt to studies such as TEEB, drawing on the latter, for example, to bolster the argument that extended conservation measures could be compatible with economic prosperity:

> The Economics of Ecosystems and Biodiversity Study shows that protected natural areas can yield returns many times higher than the cost of their protection. There are multi-million pound opportunities available from greener goods and services, and from markets that protect nature's services.[51]

Although the White Paper also announced plans to create new Nature Improvement Areas and outlined measures intended to encourage greater educational and recreational use of green spaces, a considerable number of its innovations were directed at facilitating interactions between government and businesses, on the basis of their "shared interest in protecting natural capital". Entitled *The Natural Choice: Securing the Value of Nature*, it put forward the government's intention to experiment with schemes "in which payments are made by the beneficiary of a natural service to the provider of that service" and to set up an Ecosystems Markets Taskforce with a remit to explore the possibilities for the development of 'ecosystem service' markets and trade in 'green goods'. It also announced that the

---

http://www.defra.gov.uk/news/2012/02/09/caroline-spelman-on-uk-aims-for-rio20/

49 *The Natural Choice: Securing the Value of Nature*: 36.

50 *Ibid.*, 5.

51 *Ibid.*

government would explore possible uses of "biodiversity offsets" in planning policy, by launching voluntary pilot schemes in which property developers could be involved in financing "compensatory habitat expansion or restoration" on other sites as a means of fulfilling the environmental requirements of their planning applications[52].

The adoption of 'natural capital' as a policy concept in the UK Devolved Administrations

With the exceptions of its references to UK national accounting and to international projects, *The Natural Choice* only sets out policies for England – which is unsurprising, given that responsibility for environmental issues in Northern Ireland, Scotland and Wales has largely been transferred to the Devolved Administrations, working within frameworks which include elements of devolved, UK and European legislation. Attempts to develop 'natural capital accounting' at the British level will nonetheless call for a high level of interaction between the various administrations as the process unfolds, and the White Paper is conspicuously lacking in reference to the possibilities for such exchange. However, its evidence base draws extensively on the UK National Ecosystems Assessment (the UK NEA), the preparation for which had involved collaboration between Defra, the Scottish Government, the Welsh Assembly Government and the Northern Ireland Executive[53]. The UK NEA called for the development of more integrated general environmental management but also suggested that, paradoxically, the divergence in approaches around the UK could be a positive factor in moving towards this aim:

> […] the sustainable management of biodiversity, ecosystems and ecosystem services will be made easier by using integrated approaches […] Broadly, the trends suggest that responses are becoming more integrated and reflective of ecosystem thinking, which suggests that the overall direction of change is positive. Moreover, in an international context, European Union/UK approaches to ecosystem management reflect more

[52] *Ibid.*, 22.

[53] Carried out between mid-2009 and early 2011, The UK National Ecosystems Assessment (NEA) was funded by Defra, the Scottish Government, the Welsh Assembly Government, the Northern Ireland Executive, the Natural Environment Research Council and the Economic and Social Research Council. Follow-on work is ongoing, as part of the 'Living with Environmental Change' initiative. http://uknea.unep-wcmc.org/About/tabid/56/Default.aspx

> integrated and collaborative modes of intervention. However, considerable challenges remain, and they should not be underestimated.
>
> Evidence from the national assessments (England, Northern Ireland, Scotland and Wales) demonstrates some divergence in approach, which provides useful benchmarks for a comparative assessment of policy options. In many ways, the UK context provides a 'controlled experiment' in which policies are differently implemented across the devolved administrations. There is considerable scope for innovation at country level, and for shared learning from these divergent approaches.[54]

Interest in the concept of 'natural capital' seems to have gained momentum more slowly in Northern Ireland than in Scotland and Wales[55]; by 2012 the Northern Ireland Executive had set up relatively few projects specifically evoking 'natural capital accounting' and 'ecosystem services', although its programme for government for the period 2011-2015 includes a commitment to revise the Northern Ireland Biodiversity Strategy with the aim of promoting "actions that will contribute to halting the loss of biodiversity and the degradation of ecosystem services"[56], whilst the Northern Ireland Environment Agency alluded to ecosystem valuation in identifying the creation of "a green economy that reflects the value of the environment" as one of its "strategic priorities" for 2012-2020[57].

Meanwhile, The Scottish National Party's dominant position in the Scottish Government since 2007 possibly accounts, in part, for an increasing emphasis on the economic value of the nation's natural environment in ministerial statements and policy documents, with the Scottish First Minister, Alex Salmond, asserting in 2008 that "the key to fulfilling our country's huge economic potential and to generating truly

---

[54] *The UK National Ecosystem Assessment: Synthesis of the key findings.* Cambridge, UNEP-WCMC, 2011: 55.
http://archive.defra.gov.uk/environment/natural/documents/UKNEA_SynthesisReport.pdf

[55] See Angela Watkins, 'An Introduction to Natural Environment Policy in Wales and a comparison with natural environment policies for other parts of the UK', National Assembly of Wales Publications, Research Service Briefings, Paper no. 12/018, May 2012. http://www.werh.org/documents/EnvPolicyWales.pdf

[56] Northern Ireland Department of the Environment, Programme for Government 2011-2015, Revised version, November 2012.
http://www.northernireland.gov.uk/pfg-delivery-plan-commitment-51.pdf

[57] Northern Ireland Environment Agency, *Our Passion, Our Place: NIEA Strategic Priorities 2012-2020*, 2012: 9. http://www.doeni.gov.uk

sustainable growth is to harness Scotland's stock of natural capital"[58]. A pilot Scottish Natural Capital Index providing qualitative assessments of seven broad habitat types was launched in 2011, with a subsequently refined version set to be updated annually.[59] The current Scottish government has also shown keen interest in developing the means to translate such qualitative assessments of the environment into monetary terms, with the development of tools to "build on international and national experiences and outcomes" in identifying and valuing the nation's "environmental assets, biodiversity and ecosystem services" forming one of its strategic research themes for 2011-2016[60].

Similarly, the Welsh Government is currently financing research into assessment and valuation techniques applicable to "ecosystems and their services", as part of work being undertaken for the elaboration of an evolving Natural Environment Framework, entitled *A Living Wales*[61]. Following the publication of the UK NEA and the English Natural Environment White Paper in June 2011, the Welsh Minister for Environment and Sustainable Development, John Griffiths, issued a statement stressing that Wales was actively developing its own environmental policy, while welcoming "the proposed focus on accounting for the value of the environment in UK National Accounts"[62]. The Welsh Government went on to launch *Sustaining a Living Wales*, a Green Paper consultation structured around the central proposal of moving towards "an ecosystem approach to environmental regulation and management", which it defines as being characterised by favouring a more integrated perspective in the place of current tendencies to address environmental issues separately[63]. It represents ongoing work on "improving our

[58] In his First Minister's Statement on the Scottish Government's programme for 2009, 'Taking Scotland Forward in 2009', 29 December 2008. http:// www.scotland.gov.uk/News/This-Week/Ministers-In-The- Media/081229000000
[59] *Ibid.*, 2-3.
[60] Scottish Government 'Strategic Research 2011-16, Theme 1: Ecosystem Services', Tender Document.
http://www.scotland.gov.uk/Resource/Doc/175356/0120154.pdf
[61] See http://www.wales.gov.uk/livingwales for details of the 'A Living Wales' Welsh Natural Environment Framework.
[62] Welsh Government Cabinet Statements (2011): John Griffiths, Minister for Environment and Sustainable Development, Written Statement: The Natural Environment Framework 'A Living Wales', June 2011.
http://wales.gov.uk/about/cabinet/cabinetstatements/2011/110615nat/?lang=en
[63] *Sustaining a Living Wales: a Green Paper on a new approach to natural resource management in Wales*, Welsh Government consultation document, no. WG13943, January 2012: 1.

understanding of ecosystems and their services and how we value them" as being the first stage in a development process aimed at "embedding" this new approach[64].

The authors of the Green Paper acknowledge that the 'ecosystem approach' they put forward is one which has already been promoted in international policy circles for several years and which is also evoked, to varying degrees, in policy documents produced by the other Devolved Administrations and by Defra. However, they argue that Wales is particularly well-placed to take a pioneering role in implementing policies and legislation which could see 'ecosystem approach' theory translated into practice:

> While this approach is now widely accepted in the international community, the extent to which it has been adopted at a national level as an operational approach to regulating or managing the environment is limited. This is due to the extent of existing systems and the inevitable challenges of managing multiple issues on an integrated basis. We believe that Wales has a unique opportunity to apply the ecosystem approach as an underpinning framework that will drive all parts of Government. Wales can show the way on tackling these issues because of our scale and evolving governance arrangements.[65]

The Welsh Green Paper remains tentative about the specific mechanisms which attempt to develop this policy might rely on, anticipating that the first detailed proposals will appear in a series of legislative bills planned for 2014-16[66]. *Sustaining a Living Wales* also acknowledges that seeking to develop a unified 'ecosystem approach' will call for a high level of consensus and co-ordination between the private and public sectors and across national borders[67].

Valuing natural capital: a dilemma for environmentalists?

While evaluations of 'natural capital' may involve non-monetary analyses of the type presented in Scotland's Natural Capital Index, the majority of such projects place the emphasis on generating economic

---

http://wales.gov.uk/docs/desh/consultation/120210nefgreenpaperen.pdf

[64] *Ibid.*, 6.

[65] *Ibid.*, 5.

[66] See Angela Watkins, 'An Introduction to Natural Environment Policy in Wales and a comparison with natural environment policies for other parts of the UK': 9.

[67] *Sustaining a Living Wales: a Green Paper on a new approach to natural resource management in Wales*: 29.

valuations, on the grounds that such an approach presents the most effective means of promoting environmental conservation measures. Attempts are frequently made to translate perceived intrinsic value into financial value, using assessment techniques such as surveys in which inhabitants of a particular area are questioned about their 'willingness to pay' for the benefits which they gain from their natural surroundings[68]. In 2012, Defra described the work of the Natural Capital Committee in terms which clearly illustrate this tendency to present the development of financial valuations for natural habitats and processes as a pragmatic necessity, arguing, in terms which recall those of the TEEB study, that "monetising the value of nature is a convenient way of assessing the trade-offs made in any decision by measuring the benefits and costs in a common unit."[69]

Archived accounts of parliamentary discussions on the subject suggest that the Coalition government's continuation of the work on 'natural capital accountancy' started under Labour has attracted little in the way of criticism from MPs in Westminster. Debate has centred on the means allocated for the application of such an approach rather than on its underlying principles. In July 2012, the cross-party Environment, Food and Rural Affairs Committee[70] expressed strong support for the proposals put forward in the English Natural Environment White Paper (NEWP) of the previous year but decried a lack of progress in the practical implementation of its recommendations, expressing concern that "effective mechanisms" did not seem to have been put in place to enable the development of 'natural capital accounting' on a wide scale and arguing that "valuing the natural environment must not be seen as an issue solely for Defra"[71]. The Coalition government responded by reiterating its

---

[68] See Peter Kareiva, Heather Tallis, Taylor H. Ricketts, Gretchen C. Daily, Stephen Polasky, eds., *Natural Capital: Theory and practice of mapping ecosystem services*, Oxford, Oxford University Press, 2011.

[69] http://www.defra.gov.uk/naturalcapitalcommittee/natural-capital/how-do-we-value-the-environment/

[70] EFRA is a cross-party departmental select committee of MPs, "appointed by the House of Commons to examine the expenditure, administration and policy of the Department for Environment, Food and Rural Affairs (Defra) and its associated public bodies." http://www.parliament.uk/business/committees/committees-a-z/commons-select/

[71] House of Commons Environment, Food and Rural Affairs Committee, Fourth report of session 2012-13: Natural Environment White Paper. Vol. 1, published 17 July 2012 by authority of the House of Commons, London: The Stationery Office Ltd

commitment to integrating 'natural capital accounting' into "key policy areas", arguing that the NEWP "sets the vision for the next fifty years".[72]

Meanwhile, the Green Party has also evoked the concept of 'natural capital' in recent policy documents, although in terms which explicitly recall its metaphorical use in *Small is Beautiful* - taking up E.F Schumacher's argument that modern industrial society is committing "the fundamental economic error of using capital as income"[73]**.** The Green Party member elected to the House of Commons in 2010, Caroline Lucas, has strongly criticised plans to develop further financial valuation – as opposed to metaphorical appreciation - of such 'natural capital'. Writing in *The Guardian* in response to the publication of the NEWP, Lucas argued:

> […] While the economic evaluation of the natural world might be a well-intentioned effort to convince economists and the business community of its importance, putting a price on the environment risks simply modifying it. Valuing and measuring natural capital in this way, and 'developing new markets for ecosystem services', perpetuates the idea that natural resources are simply there to be exploited.[74]

Lucas expressed particular concern about the methodological difficulties associated with the calculation of notional "equivalence" between habitats in the government's pilot biodiversity offsetting schemes. Yet, while a range of environmental and conservation groups have raised queries about the practicalities involved in such projects, few seem to have rejected the underlying principle of extended economic valuation of the natural world. Rather, a broad spectrum of UK nature conservation organisations has tended to adopt the terminology associated with the concept of 'natural capital accounting' in recent years. For example, the NEWP was welcomed by the Chief Executive of the Woodland Trust, Sue Holden, for its recognition of "woods and trees as cost-effective vehicles for delivering

(http://www.publications.parliament.uk/pa/cm201213/cmselect/cmenvfru/492/492.pdf).

[72] The Natural Environment White Paper: Government response to the Environment, Food and Rural Affairs Committee's fourth report of session 2012-1 (http://www.publications.parliament.uk/pa/cm201213/cmselect/cmenvfru/653/65304.htm).

[73] http://policy.greenparty.org.uk/en

[74]Caroline Lucas, M.P. 'Environment White Paper lacks policy, ambition and Defra's backing', *The Guardian*, 13 June 2011 (http://www.guardian.co.uk/environment/2011/jun/13/environment-white-paper-policy-ambition).

a range of services to society"[75], while the Scottish Wildlife Trust has argued that "valuing ecosystem goods and services can underline their essential role in maintaining a healthy - indeed a functioning - economy" and "brings those goods and services within economic decision-making systems"[76]. Meanwhile, Friends of the Earth organised a publicity stunt in 2012 in which giant bees presented a £1.8 billion invoice for "annual pollination services to British agriculture" to government economic advisors in Whitehall, on the basis of research commissioned by the group into the financial cost which would be incurred if British crops had to be pollinated by hand[77].

The tactic employed by Friends of the Earth in calling for a British Bee Action Plan reminds us that literal and figurative applications of the 'natural capital' concept may sometimes overlap and that, depending on the context, monetary valuations of natural processes or habitats can still be intended for metaphorical impact of a sort which does not necessarily endorse the idea of featuring the latter on any corporate or national balance sheets. For a considerable number of British conservation bodies, environmental groups and ecological science organisations, it seems, drawing attention to the financial risks associated with degradation of the natural world may be considered to be the most pragmatic means of encouraging widespread uptake of better environmental practice, even if this also implies promoting the commercial exchange of a widening range of 'green goods and services'[78]. But, given that such an approach could also result in diminished appreciation of the specificity of 'natural capital' by bolstering perceptions of the potential 'substitutability' of such processes and habitats[79], many organisations are nonetheless at pains to insist that the 'natural capital' concept should be handled with caution. For the Scottish Wildlife Trust, for example, it is "imperative that market

[75] Quoted in a BBC News article by Richard Black 'Ministers bid to stem nature loss', 7 June 2011 (http://www.bbc.co.uk/news/science-environment-13672600).

[76] The Scottish Wildlife Trust Policy on the 'Economics of ecosystem goods and services', 2010: 4 (http://scottishwildlifetrust.org.uk/docs).

[77] Friends of the Earth Press press release: 'Bees invoice Govt for £1.8 billion per year', 18 July 2012 (http://www.foe.co.uk).

[78] See, for example, W. J. Sutherland *et al.*, 'The identification of priority policy options for UK nature conservation', *Journal of Applied Ecology*, 2010, 47: 958.

[79] For an examination of the ethical issues raised in an international context by the focus on monetary evaluation of 'ecosystem services', see: Gary W. Luck, Kai M. A. Chan, Uta Eser, Erik Goméz-Baggethun, Bettina Matzdorf, Bryan Norton and Marion B. Potschin, 'Ethical considerations in on-ground applications of the ecosystem services concept.', *Bioscience*, December 2012/Vol.62 No.12, American Institute of Biological Sciences: 1020-1029.

measures should only be used as decision-support tools after non-market values have been considered"[80].

Furthermore, considerable debate exists around the methodological challenges associated with the concept: the very notion of taking stock of 'natural capital' carries the implicit assumption of a certain fixity and predictability of natural processes which is belied by the dynamic intricacy of ecological networks, as shown by the constantly evolving hypotheses of ecological science[81]. Specialists working in the field of ecological economics have argued recently that scientific modeling of such processes has gained considerably in sophistication since the publication of the Millennium Ecosystems Assessment in 2005, yet they continue to emphasise the immensity of the challenge which they are undertaking[82]. In 2011, scientific advisors behind the scenes in Westminster noted that "the lack of information about the level of natural capital required to maintain the capacity of ecosystem services at acceptable levels is a major impediment to the development of effective accounts."[83] In 2012 the Office of National Statistics (ONS) set out plans to produce an "overview of the value of natural capital within the UK" by 2015 and to work towards establishing accounts "in both monetary and non-monetary terms"[84], but acknowledged that "initially the scope, level of geographical disaggregation and accuracy of the estimates will be constrained by the

---

[80] The Scottish Wildlife Trust Policy on the 'Economics of ecosystem goods and services', 2010: 5.

[81] On the scientific and philosophical dilemmas associated with attempts to develop valuation techniques for 'ecosystem services', see, for example, Sarah Cornell 'Valuing ecosystem benefits in a dynamic world'. *Clim Res* 45, 2010: 261-272. Cornell argues that an over-dependence on monetary valuation may lead to a less precautionary approach to environmental policy and calls for greater interdisciplinary reflection to feed into the evolution of the ecosystem services concept.

[82] See Peter Kareiva, Heather Tallis, Taylor H. Ricketts, Gretchen C. Daily, Stephen Polasky, eds., *Natural Capital: Theory and practice of mapping ecosystem services*: 5.

[83] The Parliamentary Office of Science and Technology (Houses of Parliament) 'Natural Capital Accounting' Postnote 376, May 2011 (http://www.parliament.uk/briefing-papers/POST-PN-376).

[84] *Accounting for the value of nature in the UK: A roadmap for the development of natural capital accounts within the UK Environmental Accounts*. Office for National Statistics publications, London, December 2012: 3.

This plan of action for 2013-2020 was produced "in association with [...] Defra and in consultation with the Natural Capital Committee".

availability of the data"[85]. In some cases the application of monetary valuations might seem relatively straightforward – pilot projects are already underway in which water utility companies pay farmers to preserve natural 'filtration services' at source in order to cut water-processing costs downstream, for example[86]. But the identification of the providers and beneficiaries of such 'services' will rarely be so clear-cut and the choice of 'what' to value, as much as 'how' to value it is likely to be a source of debate over the coming years, given that certain forms of 'natural capital' will lend themselves more readily to identification and valuation than others.

Conclusion

Both public and private sector interest in the concept of 'natural capital' has gained momentum around the UK in recent years, particularly since the publication of the first findings of the TEEB study in 2008; however, significant obstacles remain to be overcome before some form of 'natural capital accounting' might be developed on a wide scale. The principle of attributing monetary values to 'ecosystem services' has garnered considerable support, ranging from growing private sector awareness of potential new markets to the assumption on the part of some conservation groups that such an approach is likely to offer the most immediate means of drawing greater attention to environmental issues in a variety of decision-making processes. It is therefore hardly surprising that so far there has been a relatively broad political consensus, both in Westminster and the Devolved Administrations, concerning the pertinence of attempting to develop more extensive methodological tools aimed at enabling economic valuations of 'natural capital'. However, assuming that more sophisticated assessment procedures can be elaborated despite the scientific challenges which this presents, debate is likely to be generated in the coming years around potential applications for such valuations, which could form the basis of a wider range of fiscal incentives for environmental conservation, for example, but which could equally be represented as demonstrating the auto-regulating potential of a more administratively de-regulated market.

A range of environmental policy documents produced recently in the UK conceive of 'natural capital' as including other types of capital, acknowledging that economic activity is embedded in and dependent on

---

[85] *Ibid.*, 6.
[86] See *The Natural Choice: Securing the Value of Nature*: 39.

natural processes and habitats rather than being in some way separate from them[87]. But with economic cost-benefit analyses continuing to be taken as a guiding principle in practice, the natural processes and habitats labelled as 'natural capital' could be treated as one category of capital amongst others, resulting simply in ever-increasing types of 'natural assets' and 'ecosystem services' being quantified in ways which would potentially facilitate their transfer from the public to the private domain[88]. For those with unshaken faith in the virtues of free-market economics, of course, this in itself would not necessarily be a cause for concern. Alternatively, one might also imagine the possibility that the specificity of 'natural capital' could be increasingly recognised, leading to a fundamental rethinking of human relations with the rest of the biophysical world and the refining of interactions between private and public spheres, between regions and between nations, in ways which might result in the current socio-economic paradigm evolving from within.

---

[87] See, for example, the 'Natural Capital Accounting' Postnote 376, May 2011. This research briefing observes that: "Human wellbeing arises from the use of a combination of types of capital: social capital, human capital and built capital; but these are all based on natural capital."

[88] For a neo-Marxist perspective on the issue, see Paul Burkett, *Marxism and Ecological Economics: Towards a red and green political economy,* Leidun, Boston: Brill (Historical Materialism book series), 2006.

GAIA: MYTH BEYOND SCIENCE, ETHICS BEYOND POLITICS?

CORALIE RAFFENNE
UNIVERSITY OF ORLÉANS

James Lovelock personifies the imaginary figure of the English eccentric. This might have proved a cultural obstacle to the reception of his theories in France, where he has been criticized and caricatured as a fundamentalist associated with the deep ecology movement by some mainstream intellectuals such as Luc Ferry[1] or academic scientists[2] within the prestigious CNRS. Continental skepticism with regards to deep ecology may be connected to differences in the history of ecological thought, which is rooted in botanical science and the art of taxonomy in France, and does not share the literary origins of Anglo-American ecology. The French moral philosophy tradition has been little influenced by the development of environmental ethics in the English speaking world as well as in Northern Europe, and seems at pains to broaden its disciplinary framework. However, the media exposure of deep ecology or associated activism has contributed to the inclusion of some of the ideas fostered by environmental ethics in the environmental debate throughout continental Europe.

The very British James Lovelock occupies a unique position in both political and scientific landscapes. He has benefited from wide media coverage and thus contributed to consensus building on climate change in Britain. He might be perceived as an extremist outside the UK, yet for many, he is far from radical ecology as demonstrated by his practical and professional commitments – he has worked for NASA, Shell or the secret services. His recent stance in favour of nuclear energy would appear in total contradiction with current political ecology struggles in Europe.

---

[1] *Le nouvel ordre écologique (l'arbre, l'animal et l'homme),* Paris, Grasset, 1992.
[2] J Blondel, «De l'utopie écologiste au développement durable», *Etudes*, 2003/10 Tome 399, p 327-337.

Lovelock is critical of political ecology's environmentalism which he sees as anthropocentric and essentially concerned with the preservation of the natural resources necessary to the western lifestyle. Indeed, his provocative statements could seem to embrace Naess' "biospheric egalitarianism", stressing the interdependency and equal value of all the living and non-living constituents of the Earth system. This paper does not attempt to classify Lovelock's work within the extremely broad spectrum of environmental ethics. It focuses on the most influential aspect of his theories, namely the Gaia hypothesis and explores its implications as an ethical and a political proposition in order to sketch the complex relationship between science and story-telling, politics and ethics, in ecological thinking.

The Gaia hypothesis: a double bind?

James Lovelock is known worldwide for a scientific proposition, elaborated in the 1970s with the American Lynn Margualis and known as the Gaia Hypothesis. This scientific postulate arose from the observation of atmospheric chemistry in the context of investigations on life on Mars. Contrary to what had been observed with regards to Martian atmosphere, the atmosphere of the Earth is in an extreme state of disequilibrium as a result of the multiple life forms and activities. Yet the composition of the Earth's atmosphere has remained fairly stable over geological periods of time. Hence, life regulates the Earth's atmosphere through complex interactions. The Gaia hypothesis describes the Earth as a self-regulating entity, much like a living organism, where all life-forms interact. The Earth's climate is close to optimal for dominant organisms thanks to this self-regulating capacity. In later scientific experimentations, Lovelock worked on a computer model called *Daisyworld* in order to demonstrate this hypothesis. This simplified model of life on Earth explores the interactions between two types of daisies, one black the other white, and the Sun's intensity, and their effect on the composition of the atmosphere. This self-regulation is not a permanent state and will eventually collapse in around 1 billion years. Human activity is affecting the current atmospheric conditions to an unprecedented level and will generate a transition to an alternative state, much less favorable to dominant forms of life.

The Gaia hypothesis has two dimensions which may seem in contradiction with each other:

- A scientific systemic approach

The Gaia hypothesis presents the Earth as a system and has given rise to the new discipline of Earth System Science. It proposes a global vision of Earth as a self regulated system. The essential property of this system is homeostasis, understood as the capacity to maintain its own equilibrium and present in most living organisms. The Gaia hypothesis is caught between old and new visions of ecology. Climax ecology characterises classical nineteenth century theories such as that of the Austrian geologist Suess, but also Vernadsky's global vision. In such a vision, human beings will necessarily destabilize Nature's equilibrium, which needs to be protected through limiting human presence and intervention. Recent ecology focuses on feedback and thresholds and the capacity of ecosystems to adapt to change. By focusing on the adaptability of ecosystems, human beings are ranked as one of the elements of such systems rather than an external destructive form. The Gaia hypothesis has evolved from the climax ecology theses of Vernadsky and takes into account the question of fundamental thresholds and the importance of biodiversity. Lovelock states that whilst the Earth will recover from an altered state, humanity might not: what is at stake is not the protection of Nature but the survival of civilisation.

- A metaphorical and ontological holism: Earth as a super organism

Gaia refers to the Earth as a living being capable of regulating itself in order to preserve life. Gaia is a mythical figure, the Greek goddess representing Mother Earth; this reference to one of the most ancient human symbols for Motherhood and Nature was suggested to Lovelock by the writer William Golding. Thus, the Gaia hypothesis is situated at the crossroads of diverse discursive practices and knowledge, between old and new ecology, ancient myths and futuristic prophecies of the apocalypse. It is precisely in this hybridisation of discourses that the originality and novelty of the Gaia hypothesis reside.

The Gaia metaphor: a moral fable?

According to B Latour and E Hache[3], the Gaia hypothesis illustrates a postmodern conception of morality as openness and empathy towards the

---

[3] Hache Emilie et Latour Bruno, « Morale ou moralisme ? Un exercice de sensibilisation», *Raisons politiques*, 2009/2 n° 34, p. 143-165.

rest of the world. This conception of morality, or rather ethics, is to be contrasted with the modern approach according to which moral judgment requires the establishment of a distance between the thinking unfeeling subject and the object of his/her moral preoccupations. The Gaia metaphor is a crucial element of Lovelock's ethical proposition. The metaphor stresses the living nature of Earth and thus the necessity to reconsider our relationship to it. The metaphor encourages us to feel responsibility and interdependency. The Earth can no longer be perceived as an object, even if we know that it cannot be considered as a fully fledged organism. The limits of its personification nevertheless oblige us to question ourselves on the past and present constructions of the environment. We know that Gaia is a metaphor and that it cannot constitute a totalising truth, and so we have to further question ourselves on how we understand the environment and our relation to it. The use of this metaphor is a way of engaging us in a moral questioning about the nature of our relationship to the environment, which cannot simply be ignored as a mere inanimate object. According to Latour and Hache, the use of a metaphor allows us to break free from the modern tradition of indifference and distance as a guarantee of moral objectivity. Yet, the moral fable is operative because of its anthropomorphism which is opposed and used to counteract the anthropocentrism of traditional ethical questioning. Lovelock's metaphor illustrates a strategic approach to ethics. In an anthropocentric culture, the environment has to be personified to be included in our moral framework. This illustrates the aporia of ecocentrism; value systems, even those which put Nature at their heart, will always be generated, followed and infringed by human beings. Lovelock is well aware of that and presents his moral fable as a humanistic one. The Gaia hypothesis is not about saving Earth from human destruction:

> The whole point is that the planet recovers from these events, even disasters in which 90 percent of species have been wiped out. To me, this shows how effective Gaia is at bringing the earth back from the brink of catastrophe to a healthy state. That's what homeostasis is all about. It means that a system returns to its optimum state even after it has been perturbed.[4]

It is about saving Humanity from Man:

---

[4] In James Lovelock, *The Vanishing Face of Gaia: A Final Warning* , Basic Books, London, 2009

> This is the very essence of Gaia theory…all the evidence is that Gaia will survive…even people will survive, in greatly reduced numbers. What people don't seem to get is the message that what is at risk is civilization.[5]

Thus, the Gaia metaphor remains consistent with the Enlightenment's moral and humanistic tradition.

Gaia and politics: towards ecofascism?

The Amsterdam declaration of 2001 expresses a scientific consensus which solidifies the Gaia hypothesis into a fully fledged scientific truth. It states:

> The scientific communities of four international global change research programmes - the International Geosphere-Biosphere Programme (IGBP), the International Human Dimensions Programme on Global Environmental Change (IHDP), the World Climate Research Programme (WCRP) and the international biodiversity programme DIVERSITAS - recognise that, in addition to the threat of significant climate change, there is growing concern over the ever-increasing human modification of other aspects of the global environment and the consequent implications for human well-being. Basic goods and services supplied by the planetary life support system, such as food, water, clean air and an environment conducive to human health, are being affected increasingly by global change (…) The Earth System behaves as a single, self-regulating system comprised of physical, chemical, biological and human components. The interactions and feedbacks between the component parts are complex and exhibit multi-scale temporal and spatial variability. The understanding of the natural dynamics of the Earth System has advanced greatly in recent years and provides a sound basis for evaluating the effects and consequences of human-driven change. Human activities are significantly influencing Earth's environment in many ways in addition to greenhouse gas emissions and climate change. Anthropogenic changes to Earth's land surface, oceans, coasts and atmosphere and to biological diversity, the water cycle and biogeochemical cycles are clearly identifiable beyond natural variability. They are equal to some of the great forces of nature in their extent and impact. Many are accelerating. Global change is real and is happening now. (…) Earth System dynamics are characterised by critical thresholds and abrupt changes. Human activities could inadvertently trigger such changes with severe consequences for Earth's environment and inhabitants.(…) In terms of some key environmental parameters, the Earth System has moved well outside the range of the natural variability exhibited over the last half million years at least. The nature of changes

[5] James Lovelock, *ibidem.*

now occurring simultaneously in the Earth System, their magnitudes and rates of change are unprecedented. On this basis the international global change programmes urge governments, public and private institutions and people of the world to agree that:

An ethical framework for global stewardship and strategies for Earth System management are urgently needed. The accelerating human transformation of the Earth's environment is not sustainable. Therefore, the business-as-usual way of dealing with the Earth System is not an option. It has to be replaced as soon as possible by deliberate strategies of good management that sustain the Earth's environment while meeting social and economic development objectives.

This declaration stresses the ethical dimension which remains in the scientific consecration of Gaia, but also the global scale of the problems raised and the state of emergency which has now been reached. The formulation of a scientific consensus around Gaia leaves unresolved many questions about the political implementation of measures reflecting this 'Gaian' awareness. Lovelock's vision of Gaia as revengeful reflects the heuristics of fear proposed by Hans Jonas. For the German philosopher, fear is the first component of this ethics of responsibility as it calls for action by encouraging the duty to act through fear. The heuristics of fear leads to rational humanistic perception and action. To what extent can these ethical considerations be translated into real policy and political priorities?

Climate change, ecological emergency and state of exception

What's the alternative to democracy? There isn't one. But even the best democracies agree that when a major war approaches, democracy must be put on hold for the time being. I have a feeling that climate change may be an issue as severe as a war. It may be necessary to put democracy on hold for a while.[6]

Hence, according to Lovelock, in a state of emergency, the rule of law should be suspended in order to preserve such essential values as life, a sort of green state of exception. The Italian philosopher Agamben has described how democratic states have established states of exception in times of crisis, extending forms of power beyond the original needs and objective. Even confined and justified states of exception tend to lead to

---

[6] James Lovelock , *The Guardian*, March 2010, (http://www.guardian.co.uk/environment/blog/2010/mar/29/james-lovelock).

authoritarian extensions of power and to affect rights and citizenship. During such times of extension of power, certain forms of knowledge are privileged and accepted as true, particularly if they are backed by hard science. In a green state of exception, scare-mongering ecological claims would be established as absolute truths and certain voices, including other less alarmist green views shall no longer be heard. In a time of crisis, knowledge can be violently suppressed. The neo-Malthusian slant of Lovelock's theories as well as Naess' deep ecology can only lead to severe restrictions on individual rights. There is within deep ecology's ethical discourse an inherent risk for democracy. However, it can also be argued that there are within our democracies numerous states of exception, which allow the suspension of environmental protection rules in order to facilitate the maximisation of profit.

Global scales and a self-fulfilling prophecy of doom

In the context of the disappointing outcome of numerous climate change conferences such as the 2010 Copenhagen Conference, the limits of the international rule-making process cannot be denied. The Gaia hypothesis stresses the planetary scale of environmental issues such as climate change and biodiversity. However, there is no institutional framework capable of providing global responses to planetary problems, as legal and political systems have developed on the basis of the nation-state. Lovelock would seem justified in asserting:

> We need a more authoritative world. We've become a sort of cheeky, egalitarian world where everyone can have their say. It's all very well, but there are certain circumstances – a war is a typical example – where you can't do that. You've got to have a few people with authority who you trust who are running it.[7]

Global scales call for radical solutions which do not seem capable of emerging from intergovernmental negotiation. This excludes the democratic political process as we know it, but also the market and consumer choice which encourages the poorest to aspire to same polluting lifestyle as the richest. The global awareness of environmental issues which arise inter alia from the Gaia hypothesis does not fit with the national or local scales of decision making and action; contradicts environmentalist approaches focused on local and participatory solutions made possible by smaller scales. Thus, this global systemic view of the

[7] Ibidem

environment necessarily carries with it a very pessimistic questioning of human destiny and political organisation. There is a logical connection between a systemic and global vision of environmental processes and problems, the pessimistic anticipation of ecological disasters and a call for authoritarian solutions.

Gaia's ethical project: Towards ecology of care

The Gaia hypothesis illustrates the ambivalent relationship between ethical and political values in the context of political ecology. To what extent can ecological thinking be translated into politics without a (dangerous) questioning of fundamental liberal and democratic values? The political force of ecology resides more in the self-questioning it provokes, than in any attempt to formulate it as the basis of political action, understood as state imposed objectives and rules. As such, is it bound to remain at the margins of the political system?

The Gaia hypothesis indeed pushes the subject to reconsider his/her way of being in the world, of questioning his/her identity as consumer and the nature of his/her relations to others and to the environment. It echoes deep ecology's relationalism, according to which the identity of all living things is constituted by its relations to other things in the world. This necessarily brings the individual to take better care of nature. It has been stressed that this relational vision of ecology is compatible with the notion of care developed by feminist writers such as Joan Tronto.

The philosophical work on care attempts to demonstrate that the boundaries between ethics and politics, the private and the public, can and should be transcended to create a new understanding of political organisation. This is also part the purpose of Michel Foucault in his later work on the ethics of the self.[8] As Michel Foucault has argued, the ethical self-constitution of the subject is the necessary prelude to political action. Gaia can be included in the process of self constitution of the ethical subject as a resisting subject. Hence, the Gaia hypothesis can usefully fulfill its political destiny if it is integrated in an ethical practice of freedom.

---

[8] M. Foucault, *Le Souci de soi*, Gallimard, Paris, 1984.

PART II:

DISCOURSE ET REPRESENTATIONS

A Portrait of the Expert in a Young Field

Camille Biros
University of Bordeaux II

Expertise is a word that is used more and more in our world. In the media in particular, the expert plays an essential role and is often valued above the scientist, to support ideas on all sorts of questions. Thinking about "expertise" demands a formal definition: according to *The Penguin English Dictionary* "expertise" is "skill in or knowledge of a particular field; know-how". The noun "expert" refers to somebody "having or showing special skill or knowledge derived from training or experience". What is striking is that this definition emphasises the practical character of expertise. It can be compared to "know-how" and even if it is a form of knowledge, it is knowledge of a particular field, knowledge that enables one to act in this field. To acquire expertise is not to develop theoretical knowledge for its own sake. The fact that the term "expertise" rather than "knowledge" is used more and more seems significant. It could be the sign of a shift in the way we conceive knowledge today, a shift towards a more practical approach to defining knowledge.

> Over the last half-century, the most important transformation in the way expertise has been understood is a move away from seeing knowledge and ability as quasi logical or mathematical and toward a more wisdom-based or competence-based model (Collins & Evans 2007 : 23).

Favouring practical competence over theoretical knowledge suggests a certain disdain for abstract thought. The phenomenon is particularly marked when dealing with environmental issues. Those who offer scientific analysis of problems are criticised when they fail to propose practical solutions. If the one and only aim of the environmental field is to preserve, protect or regenerate the environment, a practical approach is imperative. This may explain why the role of "experts", as opposed to scientists, politicians, and other environmental actors, is so important in the field. Furthermore, these experts often acquire a far-reaching influence.

In a field where specialization is still in a state of emergence, expertise is a rare and valuable resource and those who can show they have acquired it often benefit from it and may come to occupy a leading position in the network of environmental actors.

Considering how important expertise appears to be in the environmental field, we have set out to study the criteria that are used to establish some actors in the field as experts rather than others. Our approach is dependent on the tools developed in discourse analysis, and more particularly in the French school of discourse analysis. Our work starts with an acknowledgement of the fact there seems to be a lack of usable criteria to identify experts in the environmental field. Yet many organisations need to back up their work with informed data concerning the environment. By studying textual manifestations of intertextuality in a selected sample of publications of organisations in the UK, we try and sketch a portrait of the figure of the expert. Our sample was selected in such a way as to highlight contrasts between the publications of different types of organisations and to note an evolution over a ten years period.

Meta-criteria in the environmental field

To characterise the expert in the environmental field, one can start by determining what types of criteria are available in general to identify experts in a field. If one follows the analysis offered by Collins & Evans in the work *Rethinking Expertise*, different criteria can be noted. We base this analysis on their "periodic table of expertises" (Collins & Evans 2007: 14) and more specifically on the line entitled "meta-criteria", that is to say, "externally measured criteria". According to Collins & Evans three types of "meta-criteria" exist: "credentials" or "certificates", the "track record" and "experience". These two last criteria, "track record" and "experience" are based on past history. Thus, it is difficult to rely upon them in the environmental field that has quite a short history[1]. As regards "credentials"

[1] The question of the identification of the origin of the environmental movement is quite controversial. Many ancient texts are quoted as a proof that environmental issues have preoccupied men since times immemorial. This is not an idea we would wish to contradict but the noun itself "the environment" only started to be used as such in a mainstream context a few decades ago. The adjective "environmental" followed. Many words and concepts that are essential to the field have been created quite recently: sustainability, climate change, renewable energy, carbon neutral, to name but a few. Although a concern for nature and for mankind's surroundings has existed for a long time, the environmental field as we know it today, with its rules, stable terminology, norms, can be considered as

and "certificates", the main problem is that they don't exist in every field. In the environmental domain, specialised diplomas are still developing. There is no one particular diploma that can be considered as authoritative for the moment.

One way of finding out whether certifications and experience are considered as reliable criteria in the field is to look at environmental job offers. An environmental job website for the UK is available online and provides easy access to this type of announcement[2]. After having examined the forty last announcements published on the website on the 21st of June 2012, we analysed them to find out how many specified that the candidate had to have some sort of experience, track record or degree in the environmental field before applying for the job. We were inclusive in our choice of criteria and listed any sort of requirement linked to the field, whether it was the need for a specific certification or simply a statement that an interest, of any kind, in environmental issues would be appreciated.

The first result we got from this study was that out of forty job advertisements, six mentioned two criteria linked to the environmental field, seventeen mentioned only one, and sixteen mentioned none at all. This means that in sixteen cases, candidates need have no familiarity with the environmental field at all before applying for the job. Among the job announcements that did mention that the candidate had to have qualities that were specific to the environmental field, several specified that this was desirable but that candidates that had none could also apply. We have created a table presenting the exact figures concerning these job offers. We identified several categories in which the statements concerning the candidate can be sorted. The first element that enabled classification was to find out whether the environmental criteria were presented as a requirement or as a simple advantage. Then, we worked out whether the criteria listed were a certification, some experience in the field, knowledge of the issues that were considered as important in the field, or a simple interest in these issues with no particular knowledge to build on. The classification of the content of the forty announcements according to these criteria gives the results appearing in the following table:

---

relatively new. This is why we would call it an emerging specialised field (Biros 2011).

[2] This website named environmentjob.co.uk can be found at the following address: http://www.environmentjob.co.uk/.

Table 1

Requirements				Advantages			
Certifications	Experience	Knowledge	Taste	Certifications	Experience	Knowledge	Taste
4	9	5	2	2.5[3]	2	2	2

From this table, we can conclude that certifications do not play a very important role in selecting people designed to work in the environmental field as they are mentioned in only four job advertisements out of forty. Furthermore, in all the cases where a degree is mentioned, the way this degree is described is very vague. For instance in a job offer for a practical project officer, one can find the following description: "HND[4] or similar in habitat or heritage management, nature conservation or related subject"[5]. This tends to confirm the idea that no degree is considered as the reference. Experience seems to serve as a better indicator as it is mentioned in nine cases. Still, that remains quite a low figure. Knowledge about the issues or taste for these questions is considered as enough in seven cases. Finally, in eight cases, the environmental criteria linked to the candidate description are mentioned as an advantage but not as a requirement. Although we would need to confront the results that we have obtained here with a more in-depth study of job offers, we consider they may already point towards significant trends and show that the "meta-criteria" that are usually used to identify experts in a field can not really be relied upon here.

The expert as outlined by UK organisations

Because the "meta-criteria" receive a limited use in the environmental field, discourse analysis seems an interesting tool to advance possible understanding of the identification of experts in the field. Adopting a purely pragmatic approach, the question we can then try to answer is not the one of who *should* be considered as an expert in the field and what criteria can be used to identify him but rather who *is* considered as an

[3] In one case, the certification considered as an advantage is described in the following terms: "suitable qualification in a relevant engineering or environmental discipline"[3]. This means that an engineering degree is considered as just as appropriate as an environmental one, which is why we classified this criterion as representing 0.5 in the certification column.

[4] Higher National Diploma

[5] Http://www.environmentjob.co.uk/job/34590-Practical-Project-Officer

expert in the field. Many organisations depend on environmental information that is considered reliable to perform their tasks efficiently. In different documents that enable them to communicate with the public and to explain their strategies, they explain their views on environmental subjects and what these views are based on. One document that is particularly useful in that sense is the Corporate Social Responsibility Report. In this type of report, that was first published only by companies but is also issued by other types of organisation today, an important part is often devoted to explaining the organisation's environmental policies. In these sections, references to experts are frequent. This is the material on which we decided to base our study aiming at uncovering the mechanisms that lead some people to be considered as experts in the field. In terms of linguistic tools, intertextuality and the study of its manifestations is what will enable us to draw conclusions on the issue.

The corpus has been selected to be representative of large UK organisations. It was important to choose documents sharing the same generic features because according to the genre you communicate with, you will not use intertextuality in the same way. The Corporate Responsibility Report is a genre that is published by more and more UK large organisations and that plays an important role in the environmental field. An advantage of studying this genre is that most large companies in the UK publish a report of this type every year[6]. This is an asset when one is trying to study the evolution of certain linguistic devices over the years. The CSR reports studied are firstly those issued by a group of ten companies considered to be representative of UK large companies as a whole[7]. But although CSR reports are mainly used by companies, other

---

[6] Since the publication of the Company Law Reform Bill, "1300 companies quoted on the UK Stock Market must report on environmental matters, employees, social and community issues and risks down company supply chains" (www.publications.parliament.uk/pa/cm200506/cmbills/190/en/06190x.htm retrieved 04/11/08).

[7] To do so, we selected the company considered as the largest in financial terms in each sector identified by the London Stock Exchange and collected all the CSR reports it had published. Because we started this work in 2008, the selection of companies is based on a document published by the London Stock Exchange in 2008 entitled *Company Files 2007*. For each sector, we have selected the company having the largest income, which resulted in the following list: BAE Systems representing the Industrials sector, British American Tobacco representing the Consumer Goods sector, British Petroleum representing Oil and Gas, GlaxoSmithKline representing Health Care, HSBC representing Financials, National Grid representing Utilities, Rio Tinto representing Basic, Sage Group

types of organisation may also publish them and we found it interesting to contrast the results obtained for the companies with those obtained for an NGO and a governmental organisation. The only large UK NGO who published the equivalent of a CSR report, entitled *Environmental Report*[8], was the WWF, with reports published from 2000 onwards. The Department of Work and Pensions appeared to us as the governmental organisation having published the most CSR reports during this period, with CSR reports published every year from 2001 onwards, which is why we selected it for our study. We proceeded to a study of intertextuality in this corpus[9], separating our findings according to the type of organisation, that is to say company, NGO or governmental organisation, and the year the document was published, that is to say from 1998 to 2009.

In this corpus, we noted down all the extracts containing intertextuality. We take the term intertextuality as the French discourse analysts Charaudeau and Maingueneau define it in their discourse analysis dictionary: "la presence d'un texte dans un autre (par citation, allusion…)"[10]. All these extracts were then classified into tables where one can find synthetic information about them: document from which they were extracted, source to which they refer, theme of the extract. For example, in the 2002 BP report, one can find the following sentence:

> We support the work undertaken by individuals and many organizations, such as the World Conservation Union (IUCN), in developing a consistent

representing Technology, Tesco representing Consumer Services, Vodafone representing Telecommunications.

[8] Corporate Social Responsibility reports deal with more than environmental questions. For instance, they contain information on the social repercussions of the companies work or of those of its employment policies. Yet environmental questions are a capital part of these reports, especially if you adopt a historical point of view, as many companies started by publishing exclusively environmental information in those reports.

[9] Considering how large the corpus was, and considering the minute type of analysis we planned to perform, we couldn't possibly deal with every single document thus selected but worked from a sample of these. We considered all the reports of the NGO and the governmental organisation and all the reports of one company, BP, selected because it was the first to publish a report regularly, as early as from 1998 onwards. For the other companies, we selected one different company each year so that for each year we base our analysis on the BP report and on one other company report in such a way that each company other than BP appears once in our corpus.

[10] "The presence of a text embedded in another (through quotation, allusion…)."

> approach to the identification and designation of protected areas (BP 2002: 17).

In this extract the source is the IUCN and the theme is protected areas. The table we established with all these extracts enabled us to acquire a synthetic view of intertextuality in the different reports. From there, we were able to identify a few categories of intertextuality that are more or less well represented in different types of CSR reports. It enabled us to draw conclusions on the type of intertextuality that is most present in these reports.

A fragmented image in company reports

If we look at intertextuality in company reports, one well-established category is that of references to documents published by the organisation that is writing the report. For example, in Vodafone's 2001 CSR report, one can find the following statement: "It is a companion document to our *Report & Accounts Annual Review*" (Vodafone 2001: 3). One can also find many references to the document itself, which enable an easy navigation inside the document. Another common type of extract is that referring to external sources that made positive comments on the organisation. British American Tobacco quotes the organisation Accountability as a way of highlighting the quality of its own standards:

> All our participating Group companies are following the AA1000 Standard, a rigorous reporting standard established by the Institute of Social and Ethical Accountability (BAT 2002 : 15).

Stakeholders are also often quoted to show that they have a positive judgment on the company. In a few cases, one can find less positive comments on the organisation but usually these are accompanied by explanations tending to contest what has been said. In the 2004 Tesco report, we can find the following extract:

> In July 2003, we were identified by the Environment Agency as a 'repeat offender' following our prosecution for a petrol leak at one of our petrol stations in 2002 and a prosecution the previous year regarding the accumulation of trolleys from a store in a local river. We were surprised to be highlighted in this way, for what were two unrelated and relatively small pollution incidents (Tesco 2004 : 31).

The second sentence in this extract is a sort of justification; the organisation's judgment was not accurate according to Tesco. This tends to show that the rare negative comments on the company that are included tend to be presented as false claims.

The presence of many references to the speech of people presented as experts is also remarkable. As previously quoted, expertise on protected areas is referred to in the 2002 BP report. Another example is a reference to expertise on climate change in the BP 2007 report:

> In its Fourth Assessment Report, published in 2007, the Intergovernmental Panel on Climate Change (IPCC), established by the UN to provide a balanced view of the issue, stated that warming of the climate system was 'unequivocal' (BP 2007 : 28).

Often, companies use the people presented as experts to strengthen their own judgment. This way, they show that their practices are based on opinions that gain legitimacy through the status of their authors who are presented as experts. This is what we can observe in the following extract in which BP presents a debate about the impact of its operations on Alaska's wildlife:

> Anti-development groups express concerns about the cumulative impacts of further North Slope development. Many concerns pertain to wildlife and habitat. Numerous independent scientific studies indicate little impact on caribou, other mammals, bird life and fish from 20 years of development (BP 1998: 11).

Although two conflicting opinions are presented here, one of them is presented as lacking credibility as it is that of radical groups defending a very specific point of view: "anti-development groups". The other is presented as legitimate: "independent scientific studies". The point of view BP is most comfortable with, because it enables it to keep on expanding the territory used for oil extraction, is presented as scientific whereas the other is shown as limited.

If one looks at how intertextuality evolves, one clear change between reports from the beginning of the period under study, and reports at the end, is that in the first reports, unidentified people were often quoted, whereas at the end their identity is clearly stated in most cases. In fact, one can say that as time goes by, the expert status of the person quoted is more and more highlighted. In the 2005 BP report, one can find the following sources: "US Secretary of Labor", "International Monetary Fund", and "Deputy Executive Director of the International Energy Agency". In the 2006 GSK report the following are quoted or referred to: "GRI" (Global Reporting Initiative), "Business in the Community", "Marine Stewardship Council", and "Marine Conservation Society". These are organisations and people who benefit from a very specific status and whose words have to be taken seriously. More and more clearly referenced documents are also quoted. These may be written documents as in the following example: "In its Mobility 2030 report, the World Business Council for Sustainable Development (WBCSD) set out seven goals" (BP 2006 : 36). They may also be references to speeches: "Commissioner McCreevy, European Commissioner for the Internal Market and Services, EU Governance Conference, Brussels, January 2005" (BP 2005 : 4).

Although companies are more and more precise about whom they are quoting or referring to, it is quite striking that there is a total lack of consensus, in most areas, as to who has a claim as an expert. The reason we can conclude this is that it is very rare for different companies to refer to the same sources of expertise on environmental topics. So if our objectives in this study were to identify people and organisations that could be considered as experts for the field in a widely accepted way, we can say we have been challenged in this attempt. Yet if there is no consensus as to who may be an expert, how can we make sure that companies aren't choosing the people they quote according to one and only criterion: what they want to hear said. It would then be easy to select only the words of people who say things that show the companies in their best light and then emphasise the expert status of the person or organisation quoted as a claim for legitimacy. If we consider what we said above about intertextuality in company reports, where the most represented category is that of quotes that present the company at its best, this hypothesis seems likely.

Sparse presence in other reports

WWF's reports appear as clearly distinct from the company reports if one considers how little intertextuality is present. Among notable references, one can quote those concerning the report's generic characteristics. The following quotation from the 2000 report can be found in an equivalent form in nearly all the WWF's reports:

> We used both the *Department of the Environment, Transport and Regions' Environmental Reporting Guidelines for Company Reporting on Greenhouse Gas Emissions* and *Environmental Reporting Guidelines for Company Reporting on Waste* for this report (WWF 2000 : 3).

Other references are to international norms like the International Standards Organisation (ISO) and the Forest Stewardship Council. The few quotations and allusions to external sources in the WWF reports could be due to the status of this NGO in the environmental field. Just from being an environmental NGO, it is considered to have a legitimate voice in the environmental debate and doesn't appear to need external sources to strengthen its arguments. Research done on the new role of environmental NGOs tends to confirm this:

> [O]rganizations such as Greenpeace and FoE, once generally regarded as the radical, activist end of the environmental movement, have consolidated their status as unofficial experts in various environmental policy areas and are increasingly consulted by policy-makers (Rootes 2003 : 3).

In this context, it is quite evident that WWF does not need to quote external experts to make its claims on environmental issues.

In the reports published by the Department for Work and Pensions, extracts showing intertextuality appear in higher numbers. However, if one compares the sources to those found in the company reports, one can notice that they are much less varied. The majority of extracts are from discourses produced either by the Department itself or by other governmental organisations. Only two distinct sources are quoted: BREEAM (BRE Environmental Assessment Method) and ISO. Both of these are international norms used to regulate the practices of organisations. They aren't issued by the government and don't have a legal value yet their role is to fix rules and norms of behaviour and, in this sense, we can say that this role is quite close to that of governmental organisations. Furthermore, the reports of the Department for Work and Pensions contain very little reported speech from identified individuals. Just about all the quotations are issued from a governmental organisation. The comparison between the occurrence of intertextuality in company reports and in those of an NGO and of a governmental organisation shows that companies use much more reported speech from more diverse sources than the two other organisations. But the variety of sources in these reports is such that we have had to conclude to a certain fragmentation of the figure of the expert in CSR reports. The portrait we have offered here is quite scattered with differences that seem to be due to the status of the organisation and type of claim it aims at making.

Clearer image of the expert on climate change

There is an exception where we can see that a certain consensus does emerge as to who has a right to speak out on a certain topic. The topic we are thinking of is climate change. If we look at intertextuality on climate change in all reports from a diachronic perspective, we can say that a certain consensus progressively emerges. The first point one can notice is that from 2006 onwards, the references to external sources on climate change are more numerous. This could be linked to the fact that several important documents were published in 2006. The Stern report is often quoted, particularly in 2006 where it is the source that is mentioned the most. After 2006, the IPCC publications seem to acquire the status of main reference on this subject. What we can conclude from this is that a certain consensus seems to emerge as to who can make claims of truth concerning climate change and that the IPCC is widely acknowledged as a trustworthy organisation. Yet the IPCC has been highly criticised for its working methods and this consensual status could understandably be questioned.

The IPCC is an intergovernmental panel that describes itself as "the leading international body for the assessment of climate change" with its aim being "to provide the world with a clear scientific view on the current state of knowledge in climate change and its potential environmental and socio-economic impacts"[11]. This organisation does not produce new scientific results but delivers a supposedly complete and synthetic view of all the data published on the subject:

> It reviews and assesses the most recent scientific, technical and socio-economic information produced worldwide relevant to the understanding of climate change. It does not conduct any research nor does it monitor climate related data or parameters[12].

In December 2009 however, many of its members were implicated in what came to be known as the "Climegate Scandal". Documents and emails exchanged by members of the Climatic Research Unit at the University of East Anglia, between 1996 and 2009, were published at that time. According to many journalists, they proved that the work of these scientists had not been done in strictly scientific conditions, as there was a sort of agreement between them to provide the public with a presentation of the research on climate change in such a way as to emphasise the risks linked to it rather than to present the data in a neutral way. Members of the organisation had to defend themselves from these accusations in the media. This greatly undermined their public credibility. Later in 2010, some significant mistakes that had been published in IPCC reports were highlighted. Thus, in an article published on the 17th of January 2010, the

---

[11] Http://www.ipcc.ch/organization/organization.shtml#.T_QRpb_0cuM
[12] http://www.ipcc.ch/organization/organization.shtml#.T_QRpb_0cuM

Sunday Times exposed a mistake[13] in the data concerning the melting of the Himalaya glaciers contained in the 2007 IPCC report. An analysis of the sources on which the IPCC had based this finding showed that the information came from a 2005 report published by the WWF quoting an article from *New Scientist*, which is a scientific review for the general public. The controversies about the IPCC methodology caused all the more uproar as by 2009 it was considered as having an essential role as an expert and highly influencing heads of government, especially at the time of the Copenhagen Summit. Later on, scientific committees having assessed its working methods ruled out accusations against the institution, but nevertheless, the legitimacy of its influential position was now quite radically questioned. This issue is quite a good illustration of the fragility of the status of those who are considered experts in the environmental field. NGOs, as we said, are more and more considered as such yet their main role is not scientific. The fact that a mistake of this type was found in one of their reports is a clear illustration of that. The IPCC does present itself as having a scientific role and yet its claims of political neutrality seem difficult to maintain. Indeed, it tends to present the scientific results concerning climate change in a way that some consider as biased in favour of radical measures to fight the phenomenon. These organisations have an active role in the environmental field, which means it can be difficult for them to claim an objective stance.

To conclude, this analysis of discourse does reveal some elements of interest to someone questioning how the status of the expert is being established in the environmental field. The first is that different types of organisations use intertextuality in very different ways. NGOs seem more confident to communicate their points of view on environmental issues without calling in external experts to support their claims, whereas companies more and more highlight the extent to which they base their assertions on external expertise. So the environmental NGO can, in a way, be said to have acquired its own environmental expertise. Yet there is a lack of consensus about who precisely is acknowledged as an expert. This means that companies feel free to quote just about anyone as long as they can use the argument of "ubiquitous discrimination"[14] to show that the

---

[13] The report said that all these glaciers would have melted by 2035 or even beforehand. It was a capital mistakes as the most consensual projections plan the melting of these glaciers for a date that is more like 2300.

[14] This is the label chosen by Collins and Evans to refer to a judgment on expertise in which "the judgment turns on whether the author of a scientific claim appears to have the appropriate scientific demeanor and/or the appropriate location with the

organisation or the person referred to has a confirmed credible voice. There is just one area where this is not the case. That is climate change, where the IPCC is the main acknowledged expert. Yet we have seen that the legitimacy of the IPCC's statement has been questioned radically, on account of this organisation's role being neither exclusively scientific, nor exclusively political. Its political role means that it may have a responsibility to act in the field that would detract from the claim to scientific neutrality. Environmental problems call for action. However, the main role of science is not to prompt action. Thus today to solve environmental problems we rely on the expert but one whose role is often blurred between that of politician and scientist. This is the problem that seems entrenched in the very definition of the word "expert" because this word now implies someone who has both the capacity to act and to provide clear-cut answers:

> Telle est la forme paradoxale que prend l'intervention des scientifiques dès lors qu'ils acceptent d'exercer la fonction d'expert, situation qui les met dans l'obligation de fournir, malgré l'insuffisance de leurs savoirs, cette "connaissance de cause" que leur demandent les politiques. L'obligation de répondre inscrite dans la notion même d'expertise a donc pour conséquence que l'expertise scientifique transgresse alors inéluctablement les limites du savoir scientifique sur lequel elle se fonde[15] (Roqueplo 1997 : 20).

The environmental field is an emerging specialised field where there is a lack of clearly established meta-criteria enabling us to identify legitimate experts. My approach based on discourse analysis enables me to highlight some of the negative consequences of this situation. It seems that one way of advancing on this subject would be to trace a clearer borderline between those who have a scientific role and those who have a political role. In this context transparency in admitting the status and role of each actor in the field seems to be vital.

---

social network of scientists and/or not too much of a political and financial interest in the claim".

[15] The scientists' intervention has a paradoxical form because when they agree to play the role of the expert, they find themselves in a situation where in spite of the gaps in their own understanding, they have to produce the "full knowledge of the facts" demanded by the politicians. The need to provide answers, which is part of the definition of the notion of expertise, unavoidably leads scientific expertise to overstep the limits of the scientific knowledge on which the expertise is based.

Bibliography

Primary sources

BAE SYSTEMS, 2001, *Corporate Social Responsibility Review 2001,* London: BAE Systems

—. 2002, *Corporate Social Responsibility Report 2002,* London: BAE Systems

—. 2003, *Corporate Social Responsibility Report 2003,* London: BAE Systems

—. 2004, *Corporate Responsibility Report 2004,* London: BAE Systems

—. 2005, *Corporate Responsibility Report 2005 – Real Issues and Responsibilities,* London: BAE Systems

—. 2006, *Corporate Responsibility Report 2006 – Forward Thinking,* London: BAE Systems

—. 2007, *Corporate Responsibility Report 2007 – Gaining Momentum,* London: BAE Systems

—. 2008, *Corporate Responsibility Report 2008 – Making Progress,* London: BAE Systems

BP AMOCO, 1998, *Environmental and Social Report,* Bournemouth: BP Amoco plc

—. 1999, *BPAmocoalive - Environmental and Social Update 1999,* Bournemouth: BP Amoco plc

—. 2000, *Environmental and Social Review 2000*, Bournemouth: BP Amoco plc

BP, 2001, *Performance for all our futures*, London: BP plc

—. 2002, *Environmental and Social Review 2002*, London: BP plc

—. 2003, *Defining our Path – Sustainability Report 2003*, London: BP plc

—. 2004, *Making the Right Choices – Sustainability Report 2004*, London: BP plc

—. 2005, *Making Energy More – Sustainability Report 2005,* London: BP plc

—. 2006, *Sustainability Report 2006*, London: BP plc

—. 2007, *Sustainability Report 2007*, London: BP plc

—. 2008, *Sustainability Review 2008 – 100 years of operating at the frontiers*, London: BP plc

BRITISH AMERICAN TOBACCO, 2002, *Social Report 2001/2002*, London: British American Tocacco plc

—. 2003, *Social Report 2002/03*, London: British American Tocacco plc

—. 2004, *Social Report 2003/04,* London: British American Tocacco plc

—. 2005, *2005*, London: British American Tocacco plc

—. 2007, *Sustainability Report 2007*, London: British American Tocacco plc
DWP, 2001, *Sustainable Development, Energy and Environment - 1st Annual Report*, London: Department for Work and Pensions
—. 2002, *Sustainable Development, Energy and Environment - 2nd Annual Report*, London: Department for Work and Pensions
—. 2003, *Sustainable Development, Energy and Environment - 3rd Annual Report*, London: Department for Work and Pensions
—. 2004, *Sustainable Development, Energy and Environment - 4th Annual Report*, London: Department for Work and Pensions
—. 2005, *Sustainable Development, Energy and Environment - 5th Annual Report*, London: Department for Work and Pensions
—. 2006, *Sustainable Development – Annual Report 2006*, London: Department for Work and Pensions
—. 2007, *Sustainable Development – Annual Report 2007*, London: Department for Work and Pensions
—. 2008, *Sustainable Development – Annual Report 2008*, London: Department for Work and Pensions
GLAXOSMITHKLINE, 2002, *The Impact of Medicines - Corporate and Social Responsibility Report 2002,* Brentford (Middlesex): GlaxoSmithKline plc
—. 2003, *Making a Difference Every Day – Corporate Responsibility Report 2003,* Brentford (Middlesex): GlaxoSmithKline plc
—. 2004, *GSK Corporate Responsibility Report 2004,* Brentford (Middlesex): GlaxoSmithKline plc
—. 2005, *GSK Corporate Responsibility Report,* Brentford (Middlesex): GlaxoSmithKline plc
—. 2006, *A Human Race – GSK Corporate Responsibility Report 2006,* Brentford (Middlesex): GlaxoSmithKline plc
—. 2007, *Answering the Questions that Matter – Corporate Responsibility Report 2007,* Brentford (Middlesex): GlaxoSmithKline plc
—. 2008, *Our Reponsibility,* Brentford (Middlesex): GlaxoSmithKline plc
—. 2003, *HSBC in Society – Corporate Social Responsibility Report 2003*, London: HSBC Holdings Plc
—. 2004, *HSBC Corporate Social Responsibility Report 2004*, London: HSBC Holdings Plc
—. 2005, *HSBC Corporate Social Responsibility Report 2005,* London: HSBC Holdings Plc
—. 2006, *HSBC Corporate Responsibility Report 2006*, London: HSBC Holdings Plc
—. 2007, *HSBC Sustainability Report 2007*, London: HSBC Holdings Plc

NATIONAL GRID TRANSCO, 2003, *Operating Responsibly 2002/03*, London: National Grid Transco plc

—. 2004, *Operating Responsibly – How we performed in 2003/04*, London: National Grid Transco plc

—. 2005, *Operating Responsibly – How we performed in 2004/05*, London: National Grid Transco plc

NATIONAL GRID, 2009a, *Fiscal Year 2007-2008 Corporate Responsibility Summary Report Sustainable Growth/Responsible Practice*, London: National Grid Transco plc

RIO TINTO, 1999, *Social and Environment Report 1999*, London: Rio Tinto

—. 2000, *2000 Social and Environment Review*, London: Rio Tinto

—. 2001, *Social and Environment Performance Highlights*, London: Rio Tinto

—. 2002, *Social and Environment Review Highlights*, London: Rio Tinto

—. 2003, *2003 Social and Environment Review Highlights*, London: Rio Tinto

—. 2004, *2004 Sustainable Development Review – Meeting global needs for minerals and metals*, London: Rio Tinto

—. 2005, *2005 Sustainable Development Review – Global Commitment with Local Solutions*, London: Rio Tinto

—. 2006, *2006 Sustainable Development Review – Access to Resources, People, Land, Capital*, London: Rio Tinto

—. 2007, *2007 Annual Report*, London: Rio Tinto

SAGE UK LIMITED, 2003, *Corporate Citizenship Report for the Year 2002/03*, Newcastle upon Tyne: Sage (UK) Ltd

—. 2004, *Corporate and Social Responsibility Report*, Newcastle upon Tyne: Sage (UK) Ltd

—. 2005, *Corporate Social Responsibility Report 2004-05*, Newcastle upon Tyne: Sage (UK) Ltd

—. 2006, *Corporate Social Responsibility Report 2005-2006*, Newcastle upon Tyne: Sage (UK) Ltd

TESCO, 2002, *Corporate Social Responsibility Review 2001/02*, Cheshunt (Hertfordshire): Tesco plc

—. 2003, *Corporate Social Responsibility Review 2002/03*, Cheshunt (Hertfordshire): Tesco plc

—. 2004, *Corporate Responsibility Review 2003/04*, Cheshunt (Hertfordshire): Tesco plc

—. 2005, *Corporate Responsibility Review 2005*, Cheshunt (Hertfordshire): Tesco plc

—. 2006, *Corporate Responsibility Review 2006*, Cheshunt (Hertfordshire): Tesco plc
—. 2007, *Corporate Responsibility Review 2007*, Cheshunt (Hertfordshire): Tesco plc
—. 2008, *More than the Weekly Shop - Corporate Responsibility Review 2008*, Cheshunt (Hertfordshire): Tesco plc
—. 2009, *Corporate Responsibility Report 2009*, Cheshunt (Hertfordshire): Tesco plc
VODAFONE GROUP PLC, 2001, *Vodafone Future – Corporate Social Responsibility Report 2000-2001*, Newbury: Vodafone Group Plc
—. 2002, *The World Around Us – Corporate Social Responsibility Report 2001-2002,* Newbury: Vodafone Group Plc
—. 2003, *The World Around Us – Corporate Social Responsibility Report 2002-2003,* Newbury: Vodafone Group Plc
—. 2004, *Being a responsible business, How we're doing – Corporate Social Responsibility Report 2003-2004*, Newbury: Vodafone Group Plc
—. 2005, *We said, we have, we will – Corporate Social Responsibility Report 2004-2005,* Newbury: Vodafone Group Plc
—. 2006, *We said, we have, we will – Corporate Responsibility Report for the 2006 Financial Year*, Newbury: Vodafone Group Plc
—. 2007, *The Potential for Communications – Corporate Responsibility Report for the 2006 Financial Year*, Newbury: Vodafone Group Plc
—. 2008, *One Strategy – Corporate Responsibility Report for the year ended 31 March 2008*, Newbury: Vodafone Group Plc
WWF-UK, 2000, *Environmental Report 1999-2000 – Practicing what we preach*, Surrey: WWF-UK
—. 2001, *Environmental Report WWF-UK 2000-2001*, Surrey: WWF-UK
—. 2002, *Environmental Report 2001-2002,* Surrey: WWF-UK
—. 2003, *Practicing what we preach - Environmental Report 2002-2003*, Surrey: WWF-UK
—. 2004, *Practicing what we preach - Environmental Report 2003-2004,* Surrey: WWF-UK
—. 2005, *Environmental Report WWF-UK 2004-2005,* Surrey: WWF-UK
—. 2006, *WWF-UK Annual Review 2005-2006*, Surrey: WWF-UK
—. 2009, *Conservation, Climate Change, Sustainability - Annual Review*, Surrey: WWF-UK

Secondary sources

Beacco, J.-C., 2000, *Écritures de la science dans les médias – Les Carnets du CEDISCOR*, 6, Paris: Presses de la Sorbonne Nouvelle

—. 2004, « Trois perspectives linguistiques sur la notion de genre discursif », *Langages*, 153 (1), 109-119.

Bhatia, V., 1993, *Analysing Genre: Language Use in Professional Settings*, London: Longman.

Biros, C., 2011, "Contribution à l'étude du discours environnemental : les organisations et leurs discours au Royaume-Uni", doctoral thesis in English studies, Bordeaux 2 University.

Charaudeau, P.& Maingueneau, D., 2002, *Dictionnaire d'analyse de discours*, Paris: Seuil

Collins, H. & Evans, R., 2007, *Rethinking Expertise*, Chicago: The University of Chicago Press.

Dryzek, J. & Schlosberg, D., (eds), 2005, *Debating the Earth: the Environmental Politics Reader*, Oxford: Oxford University Press.

Flottum, K. 2010, 'A linguistic and discursive view on climate change discourse', *ASp* 58, 19-37

Foucart, S., 2010, « Le travail des experts du climat lavé de tout soupçon », *Le Monde* from the 9th of july 2010

Hammam, P. *et al.* 2002, *Discours savants, discours militants : mélange des genres*, Paris: L'Harmattan

Kuhn, T. S. [TRADUIT PAR MEYER L.], 1983 [1962], *La structure des révolutions scientifiques*, Paris: Flammarion

Nolke, H., 2009, « La polyphonie de la ScaPoLine », *La polyphonie : outil heuristique, linguistique, littéraire et culturel*, Berlin: Frank & Timme

Owen, D. (ed.), 1992, *Green Reporting – Accountancy and the challenge of the nineties*, London: Chapman & Hall

Rootes, C, (ed.), 2003, *Environmental Protest in Western Europe*, Oxford: Oxford University Press

Roqueplo, P., 1997, *Entre savoir et décision, l'expertise scientifique*, Paris: INR

'HOW'S THE ENVIRONMENT?' THE REPRESENTATION OF THE ENVIRONMENTAL DEBATE IN THE TELEVISION SERIES *YES, MINISTER*

PIERRE-FRANÇOIS PEIRANO
UNIVERSITY OF TOULON, LERMA
RESEARCH LABORATORY

The quotation inserted in the title of this article is taken from a dialogue in one of the *Yes, Minister* episodes, "The Middle Class Rip-Off" (season 3, episode 7), in which Sir Humphrey Appleby (Nigel Hawthorne), the Permanent Secretary of the Department of Administrative Affairs, has a conversation with his counterpart in the Department of the Environment, during an interval at the Royal Opera House. While Sir Humphrey, just to engage in conversation, asks him a trivial question, his interlocutor thinks that Sir Humphrey is referring to the environment of the room they are in and answers: "A bit smoky, isn't it?", as if he had not noticed the reference to his occupation and was not concerned by his department.. This example is already a revealing one considering the representation of the environmental debate and issues in this series but, before going further, it is necessary to dwell on the context and to recall the importance of such a debate in the Great Britain of the early 1980s.

Probably influenced by the American parties and lobbies, the main British political parties attempted to give the environment and ecology a key role in their respective programmes, to such a point that Cambridge Professor Susan Owens entitled one of her articles, published in 1986, "Environmental politics in Britain: new paradigm or placebo?"[1] The various quotations inserted before the introduction are slogans or extracts from speeches of the time, already shedding light on the newly acquired role of the environment in the political debate. While one of the slogans of

[1] Susan Owens, "Environmental politics in Britain: new paradigm or placebo?", *in Area*, volume 18, n°3 (September 1986), pp. 195-201.

the Conservative Party was "If you are a conservationist, you should also be a Conservative", Labour Party leader Michael Foot claimed, in 1983: "The Labour Party is the only party truly committed to conserving our natural environment". Lastly, a slogan of the Liberal Party confirmed this trend: "The Liberal Party is an environmentalist party". This awareness about this issue may be accounted for by various factors:

- the consequences of a new wave of industrialisation brought about in the 1970s and partly due to Britain's entry into the Common Market, which, in turn, gave rise to a flurry of local conservationist and environmentalist movements, still active in the 1980s:

> Agriculture in particular thrived in the 1970s. Britain's entry into the Common Market seemed to portend a further advance. [...] There was a growing tendency for farming to be taken over by inorganic, mass-production techniques, [...] along with the introduction of harmful chemical methods of fertilization and soil nourishment. The massive growth of conservationist and environmentalist movements, seen in the growth of the Council for the Preservation of Rural England (and Wales), and the rapid development of the National Trust, to protect historic buildings and landscapes, was in part a voluntary protest at the very prosperity and advance of agricultural Britain.[2]

- the consequence of industrial disasters that had occurred in the course of the preceding years – like Seveso, in Italy, in July 1976, an event mentioned in one of the *Yes, Minister* episodes;
- the debate on the harmful consequences of nuclear weapons, much acute throughout the Cold War period.

As a mirror effect, this environmental debate was also given a substantial role in British television series. In the 1970s, *The Good Life* staged two couples, the Goods and the Leadbetters, living in neighbouring houses. While the former decided to reject the consumer society and desperately indulged in ecology, the latter were shown as traditional, snobbish and self-centred consumers. A few years later, this debate would feature again, though treated differently, in *Yes, Minister*, a British television series that was quite popular at the time and even received several BAFTA Awards. Broadcast on BBC Two between 1980 and 1984—and even followed by a sequel, *Yes, Prime Minister*, which will not be studied in this article—this political satire was written by Antony Jay and Jonathan Lynn, two former Cambridge students, and initially followed the political career of James Hacker (Paul Eddington). After his party wins

---

[2] Kenneth O. Morgan, *Britain since 1945: The People's Peace*, Third Edition. Oxford University Press, 2001, p. 394.

the general election, Hacker is appointed as the Minister for the Department of Administrative Affairs and, at his department, has to work with his Permanent Secretary, Sir Humphrey Appleby, as well as his Private Secretary, Bernard Woolley (Derek Fowlds). However, Sir Humphrey's interests seldom coincide with Hacker's. While the latter has to make sure he remains popular and that he carries out the decisions of the Prime Minister, the former—who actually epitomises the opinions and the thought pattern of the Civil Service—only seems to be interested in securing the position of his department in terms of staff and budget, as well as striving to block all the reforms that could upset the established order. More than a satire of the political world and the alleged incompetence of the persons in charge, the influence of the main political events of the time is reflected in *Yes, Minister*, concealing a subtle reflection on the nature of political power and the main issues characterising "post-industrial" societies, some of them tackled by the Thatcher Government in the early 1980s: unemployment, the role of the State, the cutting of public expenses—as well the environmental debate. The vague name chosen for the ministry, "Administrative Affairs", leaves room to a variety of issues susceptible of falling within its realm, as if Hacker cared about everything and nothing at the same time. Out of the twenty-two episodes of the series—including the 1984 Christmas Special "Party Games", at the end of which Hacker, against all odds, becomes the Prime Minister of Great Britain—three of them are directly concerned with the environmental debate:

- "The Right to Know" (series 1, episode 6), in which Hacker must take the decision of maintaining or cancelling the protective status of a forest supposed to provide a habitat for a colony of badgers. This episode remains quite simplistic and less attention will be paid to it;
- "The Greasy Pole" (series 2, episode 4), in which the chairman of the British Chemical Corporation wishes to produce, with Sir Humphrey's approval, a compound called "propanol" at a chemical factory on Merseyside, although it could potentially damage the environment. Nevertheless, Hacker must be persuaded that the compound is totally harmless to give the green light to the project;
- lastly, "The Quality of Life" (series 2, episode 6), in which Hacker visits a city farm threatened to be closed because of a Machiavellian plot woven by Sir Humphrey—unknown to the minister, of course.

Apart from those episodes, other references to the environment are to be found in various other ones and, after those summaries, it already appears that the environmental debate is used by politicians and top-notch civil servants without being taken very seriously—in various ways, though. Numerous examples will be taken—often from "The Quality of Life", one of the best-crafted episodes in the series—and the environment may either be considered as a secondary issue by the main characters; used as a mere pretext in the purpose of seeking popularity; or, quite simply, ridiculed.

Relegating the environment in the background as a secondary issue becomes obvious when Sir Humphrey ironically asks Bernard to write down hacker's words, as the latter is hostile to the further erection of high-rise buildings and seems to care about securing the future of his fellow citizens. In this way, it appears that Sir Humphrey is not really concerned with the consequences on the environment:

> HACKER. – *(Puts his glasses on)* Oh, yes, so I did. What does he want?
> SIR HUMPHREY. – Well, he wants to talk to you about his forthcoming application to put some more storeys on his bank's proposed new office block.
> HACKER. – Ah! Now, that's just the thing we've got to put a stop to, Humphrey. Somebody has got to speak out about the environment, take a courageous stand on high buildings!
> BERNARD. – Oh, that would make a very good press photograph, Minister. *(Gesturing the picture with his spread-out hands)*
> SIR HUMPHREY. – Yes, thank you, Bernard. Well, it's up to you, Minister, but, of course, he does have quite a good case. Have you read the background papers?
> HACKER. – *(Embarrassed)* Oh, yes, of course. Remind me.
> SIR HUMPHREY. – Well, the bank points out that there are already several tower blocks in the area and, of course, it is the international division of the bank and expanding rapidly, and banks need a central location. Of course, it would bring in a lot of great revenue for the City Council.
> HACKER. – Yes. Isn't that typical? Money, money, money… What about *beauty*, Humphrey?
> SIR HUMPHREY. – Oh, yes, indeed, quite, Minister, beauty. Make a note of that, would you, Bernard?
> HACKER. – What about our children? And our children's children?
> SIR HUMPHREY. – Exactly, Minister—"children's children", Bernard.
> *(The Quality of Life)*

The opinions of other characters, like Sir Frank Gordon, the Permanent Secretary of the Treasury, may vary dramatically. At lunch with Sir Humphrey, he begins to praise city farms—only to change his mind a few seconds later, when he is told that the closing of one of them will be profitable to his department:

SIR FRANK. – Incidentally, I saw your chap on the television last night, cuddling a rabbit.
SIR HUMPHREY. – Hmm… Saint Francis of Tower Hamlets.
SIR FRANK. – What was it supposed to be in aid of?
SIR HUMPHREY. – After the rodent vote, I imagine.
SIR FRANK. – Quite a good idea all the same, those city farms.
[…]
SIR FRANK. – Oh, what's the site being used for at the moment?
SIR HUMPHREY. – Oh, nothing of any significance. A city farm for school children.
SIR FRANK. – Like the one Saint Francis was visiting? Not *the* one? *(Sir Humphrey does not answer, meaning "yes")* Mind you, I've always said those things were a bad idea—unhygienic, a danger to public health.

(The Quality of Life)

In the first scene of *The Greasy Pole*, Sir Humphrey is talking to Sir Wally McFarlane, the chairman of the British Chemical Corporation, and both men are worried about not mentioning the name of "dioxin", which could give way to further debate on the environment and thwart their project. It is thus necessary to conceal—both to politicians and to the nation—the potential danger of the compound, as making profits is preferable to preserving the environment:

SIR HUMPHREY. – But you say propanol contains metadioxin.
SIR WALLY. – So what?
SIR HUMPHREY. – Dioxin was the chemical released at Seveso. It can damage the foetus.
SIR WALLY. – That's dioxin, for Christ's sake! Metadioxin is an inert compound. It has been given a clean bill of health by the FDA[3] in Washington. The Henderson Committee, which *you* set up, is going to give it the OK, too.
SIR HUMPHREY. – Yes, I know, Wally, I know, but the name could be, well, politically sensitive.
SIR WALLY. – But, anyway, I've taken the word "metadioxin" out of the proposal. I just call it "propanol".
SIR HUMPHREY. – Ah, very wise of you.
SIR WALLY. – You told me to.
SIR HUMPHREY. – Very wise of me.

(The Greasy Pole)

The environmental debate may also be used as a pretext—by Hacker, for example. One of his press officers, Bill Pritchard (Antony Carrick),

[3] FDA: Food and Drug Administration.

considers that the image of a minister close to the rural world and concerned with preserving the environment could increase his popularity:

BILL. – Minister, it's really excellent: "Hacker speaks out against tower blocks".
HACKER. – Yes, that's not bad.
BILL. – "Minister's courageous stand on high buildings".
HACKER. – I sound like a steeplejack… That's just the heavies! What about the popular press? How am I going to get into the mass circulation dailies?
[…]
BILL. – Otherwise, something with cuddly animals or small children—like this afternoon's visit to the city farm.
HACKER. – Ah, yes! Is everything laid on for that?
BILL. – Yes, Minister.
HACKER. – *Mail*, *Mirror*, *Express*?
BILL. – Yes. And the *Sun*, and *Today*, and *Nationwide*.[4]
HACKER. – *(Pleased) Nationwide*? I'm going to be on telly?
BILL. – With any luck.
HACKER. – Well done, Bill! That's more like it.

(The Quality of Life)

Demagoguery is not totally absent from the picture. Later, in the same episode, Hacker is relishing the publicity he obtained after his visit to the city farm while skimming through the evening papers:

BILL. – The evening paper, Minister.
HACKER. – Ah!
BILL. – *(Showing him the article)* Centre page.
HACKER. – Oh, I say, double-page spread, what about that!
BILL. – Yes, Minister.
HACKER. – *(Laughing)* Ah, "The Saint Francis of Tower Hamlets"! Ah, ah, ah, ah…
BILL. – Yes, Minister.
HACKER. – Oh, there's a picture of me with some ducks. Another with a small girl. Oh, a black… ethnic… multi… you know, one of… one of the… What a good publicity, eh?
BILL. – Yes, Minister!

(The Quality of Life)

One of the essential aims in the representation of the environmental debate in the series is to ridicule it, first through puns or witty dialogues.

[4] *Nationwide*: a BBC television programme with television reporter Sue Lawley.

Let us hear what Sir Desmond Glazebrook, a City banker, has to object to Hacker's reluctance about the erection of yet another high-rise building:

HACKER. – What about beauty?
SIR DESMOND. – Beauty? This is a building, not an oil painting.
HACKER. – And the environment?
SIR DESMOND. – Oh, yes. I promise we will make sure it's part of the environment. It's bound to be once it's there, isn't it?

(The Quality of Life)

The partisans of the preservation of nature are equally caricatured by Sir Humphrey:

HACKER. – Good, good. You don't think I ought to take it to Cabinet?
SIR HUMPHREY. – Well, in my opinion, the less said, the better.
HACKER. – Oh? Why?
SIR HUMPHREY. – Because, although metadioxin is totally harmless, the name may cause anxiety in ignorant minds.
HACKER. – Humphrey, you're talking about my Cabinet colleagues!
SIR HUMPHREY. – I was referring to the Friends of the Earth and other pressure groups.

(The Greasy Pole)

Even when Hacker is ready to make a speech at the city farm to emphasise the preservation of the environment as a capital issue, an unexpected incident ruins it all. The environmental debate even becomes a topic among others and the minister's words seem quite commonplace:

HACKER. – It's a very great pleasure to be here with you all today. *(Glancing at the paper)* You know, things are changing fast. We live in a world of change. The silicon chip is changing our lives. The quality of life is becoming more and more important. The environment, conservation, problems of pollution, the future of our children, and our children's children—these are today's issues. There is, quite rightly, an increasing concern about high-rise buildings. I'm happy to be able to reassure all of you, who are members of the Architectural Association… *(Realises that he does not have the right speech)* I'm so sorry. *(Aside, to Bernard)* It's yesterday's speech!
BERNARD. – Well, that's the one *I* had, Minister.
HACKER. – No, no, you've got it, come on.
BERNARD. – Look in your inside pocket.
HACKER. – *(Searching his pocket)* You didn't…
BERNARD. – Please…
HACKER. – *(Finds the paper in his pocket)* Oh, yes! *(Back to the microphone)* I'm so sorry! I made a speech yesterday and… anyway, this

is the one. It's a great pleasure to be here with you all at the city farm today. You know, things are changing fast. We live in a world of change. The silicon chip… *(Realises these are the same sentences)* is changing our lives. *(Mumbling)* The quality of life is becoming more and more important… the environment… conservation… pollution… future of our children's children… these are today's issues. *(Resuming the appropriate tone)* The city farm movement is a welcome and important addition to the way of life for children in inner cities and we, in the Government, feel that they have a vital part to play in our children's educational and social life. And we shall do all we can to help this movement to flourish. *(Applause)* Happy birthday!
(The Quality of Life)

Lastly, one of Sir Humphrey's answers in *A Question of Loyalty* (season 2, episode 7) will round up those examples, as he tries to justify, in front of a select committee, the costly building of a roof garden on top of the supplementary benefits office in the town of Kettering:

MRS OLDHAM. – Well, let's come to the roof garden!
SIR HUMPHREY. – Yes, with pleasure. This was a part of a wide variety of roof insulation schemes, which the Government was testing in the interest of fuel economy.
MRS OLDHAM. – But seventy-five thousand pounds!
SIR HUMPHREY. – Well, it was thought that the sale of flowers and vegetable produce might offset the cost.
MRS OLDHAM. – And did it?
SIR HUMPHREY. – Hmm, no.
MRS OLDHAM. – Then, why not abandon the garden?
SIR HUMPHREY. – Well, it's there now and… it does insulate the roof and we aren't building any more.
MRS OLDHAM. – But you wasted seventy-five thousand pounds!
SIR HUMPHREY. – Well, it was Government policy to test all proposals for fuel saving.
MRS OLDHAM. – At this fantastic waste of the taxpayers' money? You agree the money was wasted?
SIR HUMPHREY. – It's not for me to comment on Government policy. You must ask the Minister.
(A Question of Loyalty)

Through those examples, it thus appears that the environmental debate is given quite a rough ride in this series. In order to stress its verisimilitude, many references to contemporary events are made, like the Seveso chemical disaster or the anti-nuclear missile groups of women and children protesting outside the military base of Greenham Common. The typical pattern, however, is that Sir Humphrey strives to manipulate Hacker or to conceal the real stakes of an issue while making skilful moves behind his back. Hacker, for his part, wishes to use the environmental issue to increase his popularity—which does not prevent him from overlooking this subject, when it suits him. In "The Greasy Pole", he thus tries to convince Joan Littler, MP for Liverpool Southwest, that producing propanol in a Merseyside plant would bring in profits for the constituency and that the possible harmful consequences would be negligible—as the votes of "real" environmentalists do not matter anyhow:

> LITTLER. – That's not the point!
> HACKER. – What is the point?
> LITTLER. – The point is that this factory is in my constituency!
> HACKER. – But it will be good for the constituency—more jobs, more money. The only people we can possibly offend will be a few cranky environmentalists. At most, I should say, it couldn't cost us more than… about a hundred votes.
> LITTLER. – My majority is ninety-one.
>
> *(The Greasy Pole)*

So even if environmentalists are portrayed as "cranks", the following power relationships and the image could be conveyed to the spectator:
- the people already seems aware of the environmental debate and seems to care about the preservation of nature;
- for the Civil Service—and, particularly, top-notch civil servants—money profits or the "good" of the country come before ecology; the voice of the British people, "ill-guided" and "ill-educated" as Sir Humphrey describes it, does not really seem to matter;
- the Minister—and, on a larger scale, politicians—seem to be caught in the crossfire, often obliged to yield to administrative orders, but nonetheless showing political courage occasionally.

This could be triggered by private reasons: in "The Right To Know", for instance, Hacker's daughter threatens him with undressing in public in order to protest against the Government's decision. But Hacker can sometimes be more cunning than Sir Humphrey—for example, at the end of "The Greasy Pole", when he talks a Cambridge professor into changing the contents of his report and questioning the production of propanol, before enjoying the benefits of his decision not to give the green light to the British Chemical Corporation:

> *PRESENTER. – The Minister for Administrative Affairs, James Hacker, has announced that he will* not *be giving his approval for the British Chemical Corporation to manufacture propanol. The report of the Henderson Committee, while generally approving the drug, said it would*

be irresponsible to deny that future research might disclose health risks. Mr Hacker made his announcement today while on a visit to the BCC's Merseyside factory, where he was greeted by an enthusiastic crowd. And now, sport. Following last night's overwhelming defeat…
(Hacker switches the television off)
SIR HUMPHREY. – Well, Minister?
HACKER. – Well, Humphrey?
SIR HUMPHREY. – Do you feel like a hero?
HACKER. – Indeed, I do. And Number Ten will be delighted.
SIR HUMPHREY. – One of the worst governmental decisions I've *ever* witnessed!
HACKER. – And one of the best political decisions I've ever made.
(The Greasy Pole)

However, this example stands as an exception and pessimism underlies the satire. This series is only a representation, of course, but it would now be relevant to wonder whether its premise was in keeping with the reality of the time or gave a critical vision of the environmental debate.

The hypothesis that the authors, Antony Jay and Jonathan Lynn, stated openly what everyone—the British people and the politicians alike—thought but did not dare say in public, could be raised. Indeed, they never seem to side with a particular party: neither the Labour Party nor the Conservatives are mentioned in the series and the spectator does not know which party Hacker belongs to. The criticism of the political world at large thus seems to be their main goal—the first way of implying that, in the realm of ecology, the parties' line of conduct remained the same: an apparent concern, concealing indifference.

The cornerstone of the series, however, seems to be a reversal of Hegel's "Master-slave Dialectic", exposed in the *Phenomenology of Spirit* (1807). Normally, politicians are thought to be the actual experts and the masters in their own houses, but the series turns this common view upside down. In the case of the fictitious Department of Administrative Affairs, Sir Humphrey is more cunning and more intelligent than Hacker, his supposed master—often naïve, but who, in the course of the series, nonetheless learns numerous Civil Service tricks, but constantly questions the civil servants' thought pattern, according to which "everything has to change so that nothing changes".[5] One thing leading to another, it is very unlikely that the environmental issue, with its far-reaching questioning of

[5] This quotation is taken from the novel *The Leopard* (1957), by Giuseppe Tomasi di Lampedusa.

the implemented policies, will be properly tackled and the series conveys a true-to-life account of the discrepancy between the theoretical aims of politicians and civil servants, apparently attached to preserving the environment, and the implementation of those ideas. Hacker even finds himself trapped when, in "The Quality of Life", he promises that the lease of the city farm is going to be renewed and, then, that the farm is going to be closed. The warden of the farm, Mrs Phillips, burst into the Minister's office:

> MRS PHILLIPS. – *(In the private office)* I don't care if it is the Queen *and* the Pope! *(She bursts into the room) (To Hacker)* Judas!
> HACKER. – I say, look, here!
> MRS PHILLIPS. – You promised to support us!
> HACKER. – *(Mumbling)* Well, and so I did…
> MRS PHILLIPS. – Then, you must see that our lease is renewed.
> SIR HUMPHREY. – Well, dear Lady, I'm afraid it's not within the Minister's p… *(Mrs Phillips turns around and faces him. He stops immediately)* I'm sorry.
> MRS PHILLIPS. – Mr Hacker, you have given your word.
> HACKER. – Yeah, I…
> MRS PHILLIPS. – *(Interrupting him) Are* you going to keep it?
> HACKER. – Well, of course, I shall leave no st… no stone… every avenue, er… everything done that is humanly possible…
> MRS PHILLIPS. – You mean "no".
> HACKER. – No—I mean… yes—I mean… no—no, yes, I mean… er, well, definitely not! No. Yes—I mean no.
>
> *(The Quality of Life)*

In this way, the series denounces hypocrisy as the king's fool would do in a Shakespeare play—as the saying goes, "many a truth is spoken in jest". This gives more credit to Susan Owens's theory that the importance of the environmental debate in politics in the 1980s was but an illusion, a pretext to ease the politicians' conscience.

However, in this representation, the way the spectators could identify to such and such character is not to be overlooked. It would hardly be exaggerated to state that Sir Humphrey plays the part of the "villain"—a sophisticated Machiavellian—while Hacker is rather portrayed as the "good guy", a naïve politician lacking experience, but striving to take decisions, with more or less good faith, in the interest of the people. Bearing in mind such a bias and Hacker's often sincere concern for the environment—more sincere than Sir Humphrey's—its preservation is indirectly considered as profitable and necessary. Could it be, beyond the satire of all the parties that had allegedly given the priority to ecology, an indirect way to call for real action on the part of the British Government, whatever the party in power?

This hypothesis is worth being raised as well, even though it is quite difficult to bring a definitive answer. *Yes, Minister* was broadcast on the BBC, a public channel that was, at the time, under the control of the Government and could help, despite dismissing the main parties without siding with any, convey the following message: "the Government or the ministry represented in the series do not really act, but this series gives a stereotypical vision of politics, that the citizens do not approve of. So, in real life, changes will occur". Margaret Thatcher herself had confessed she had been very interested in the series, which had given her "hours of pure joy".[6] The decision taken by her successive Cabinets in the realm of ecology were even underlined by *The Economist* a few years later, in a 1989 article entitled "The Greening of Margaret Thatcher",[7] but, with hindsight, the various measures taken at the time could also seem quite superficial:

> The 'green' argument advanced by environmentalists, fanned by the importance of themes such as land conservation, acid rain, and the protection of the ozone layer and the 'greenhouse effect' in the 1989 Euro-elections, suggested that the message of Thatcherite consumption and wealth creation failed to mesh with wider concerns for the quality of life.[8]

To conclude—and widen the scope the representation of the environmental debate in the *Yes, Minister* series, less naïve than it seems, calls for awareness on the part of politicians. Even though the series is set in Great Britain, it would be possible to claim that this representation applies to most "post-industrial" countries. Many of the issues raised in the series are still acute nowadays and the authors had already sensed the power relationships between industrialists, the Civil Service and politicians. The debate on the environment proves to be quite disturbing, as it implies deep change in the economic and political models, as Timothy Doyle and Doug McEachern stated in 2001, as the importance of the business sector is indeed stressed in the series: "The important question remains: under what conditions, in what circumstances and to what extent will business seek to harmonise its profit-seeking objective with protecting and improving the environmental condition of the

---

[6] Michael Cockerell, *Live From Number 10: The Inside Story of Prime Ministers and Television*. Faber and Faber, London, 1988, p. 288.

[7] "The Greening of Margaret Thatcher", *in The Economist*, March 11, 1989.

[8] Kennet O. Morgan, *op. cit.*, p. 502.

world?"[9] To end on a good note, the hesitant or ambiguous attitude of executives of all kinds could herald the changes brought about by the coming of a new era, called "the ecological era" by Donald Worster in his compelling work *Nature's Economy: A History of Ecological Ideas.* We are perhaps in the early stages of the "new paradigm" mentioned by Susan Owens and only hindsight will probably tell the real significance of the environmental debate and of the decisions taken in this field. Its importance in the *Yes, Minister* series undeniably accounts for the role it had already been given at the time. Even though it provides comic relief, the environmental debate is nonetheless overwhelmingly present in the series and it affects numerous other domains, as shown in the various extracts transcribed in this article.

Bibliography

Works cited

Cockerell, Michael, *Live From Number 10: The Inside Story of Prime Ministers and Television*. Faber and Faber, London, 1988.
Doyle, Timothy & McEachern, Doug, *Environment and Politics*, Third Edition. Routledge, London-New York, 2008.
Morgan, Kenneth O., *Britain since 1945: The People's Peace*, Third Edition. Oxford University Press, 2001.
Worster, Donald, *Nature's Economy: A History of Ecological Ideas*, Second Edition. Cambridge University Press, 1994.

Articles cited

Owens, Susan, "Environmental politics in Britain: new paradigm or placebo?", *in Area*, volume 18, n°3 (September 1986), pp. 195-201. Internet link: http://www.jstor.org/stable/20002340 <accessed on September 12, 2012>
"The Greening of Margaret Thatcher", *in The Economist*, March 11, 1989.

Television series cited

The Good Life. BBC, 1975-78.
Yes, Minister. BBC, 1980-84.

[9] Timothy Doyle and Doug McEachern, *Environment and Politics*, Third Edition. Routledge, London-New York, 2008, p. 220.

PART III:

THE POLITICAL DEBATE IN THE UNITED-KINGDOM

The Party Politicisation of Climate and Energy Policy in Britain

Neil Carter
University of York

This chapter examines how the party politics of climate change in Britain was transformed during 2006-2007.[1] In an article on British environmental politics completed soon after the 2005 general election I concluded that "the party politicisation of the environment in Britain is limited...While no party can afford to ignore the environment, the Conservative and Labour parties have both pursued a strategy of preference-accommodation, characterised by a reactive approach to public opinion, but resisting competition over the environment" (Carter 2006: 763-764). Yet within months of writing these words the political salience of the environment, and particularly climate change, had risen rapidly up the domestic agenda and for the first time in British politics all three parties were competing aggressively on these issues.

This politicisation of climate politics was very significant. There is a longstanding assumption that the limited party politicisation of the environment in Britain (Carter 2006), and elsewhere (Weale *et al* 2000: 246-256), has contributed to the weakness of domestic environmental policies. The logic is simple: if parties are competing with each other politically on environmental issues then they will attack their rivals for weaknesses in their policies and advocate tougher measures themselves, ultimately leading the governing party (or parties) to introduce more progressive policies. Certainly one obstacle to firm action by Governments on climate change is the apparent lack of commitment amongst political elites to an issue that traditionally has low salience and requires policies that risk serious political damage (Compston and Bailey 2012). Specifically, the benefits of climate mitigation policies are uncertain,

[1] I am very grateful to the Leverhulme Trust for awarding me a Research Fellowship (award number R1337901) during 2011-2012 to enable me to work on this project.

whilst the costs may be very high because of the potential negative impact on competitiveness, jobs, energy costs and lifestyles. In short, there appears to be few short term winners and many losers. Consequently, there is little political incentive for parties to support progressive domestic climate mitigation policies.

Yet 2006 marked the start of a period of radical transformation in UK climate and energy policy that arguably outlived the defeat of the Labour Government in the 2010 general election, continuing into the first eighteen months of the Conservative-led Coalition Government. This step-change under Labour was marked by the ground breaking *Climate Change Act 2008*, which established tough 2020 and 2050 emission reduction targets on a statutory basis, five-yearly carbon budgets and an independent Climate Change Committee with a remit to advise the Government on the policies needed to achieve these targets. The legislation was backed by an innovative Low Carbon Transition Plan, an impressive array of policy measures on renewable energy, feed-in tariffs, carbon capture and storage, infrastructure planning and domestic energy efficiency, and supported by significant public investment. The Government also declared its commitment to the construction of a new generation of nuclear reactors. While Labour had developed the overall low carbon strategy and policy frameworks, considerable implementation challenges remained. The succeeding Coalition Government introduced a Green Deal on domestic energy efficiency, a Green Investment Bank and an Energy Bill to reform electricity markets by providing a supporting regulatory and financial framework, including stable and predictable incentives, to encourage companies to invest long term in low-carbon generation. Of course, plenty of gaps remain (notably in reducing emissions from transport) and much work is still needed, but there is no doubt that UK climate energy policy has undergone a major transformation since 2012.

So we have two puzzles. First, how do we explain the sudden party politicisation of climate politics from 2006? Second, to what extent did this party politicisation contribute to the step-change in climate and energy policy that started in mid-2006? This chapter addresses these questions by drawing on extensive empirical research, including interviews with over 30 elite actors (ministers, political and scientific advisors, senior civil servants and ENGO representatives) in the Labour Government and wider climate policy community during this period.

Pre-2006: The environment as a mid-term Issue

Prior to 2006, the UK had experienced only a limited party politicisation of the environment. From the mid-1980s the three major UK parties responded to growing public concern about the environment with a strategy of preference accommodation: each party gradually developed a modest range of environmental policies and adopted a greener rhetoric, but they did not try seriously to make it a subject of party competition. Something akin to a cyclical pattern emerged, with parties notably receptive to environmental concerns at the mid-term stage of the electoral cycle, as illustrated by separate flurries of policy documents from each party in 1985/6, 1990 and 1994, only for interest to subside as the next general election approached (Carter 2006: 758). Although Labour was consistently marginally greener than the Conservatives, as indicated by analysis of data from the Manifesto Project (see Figure 1),[2] the consistently low scores for both parties indicated their continuing commitment to a preference accommodation strategy.

Figure 1. Environmental protection in British party manifestos 1987 – 2010

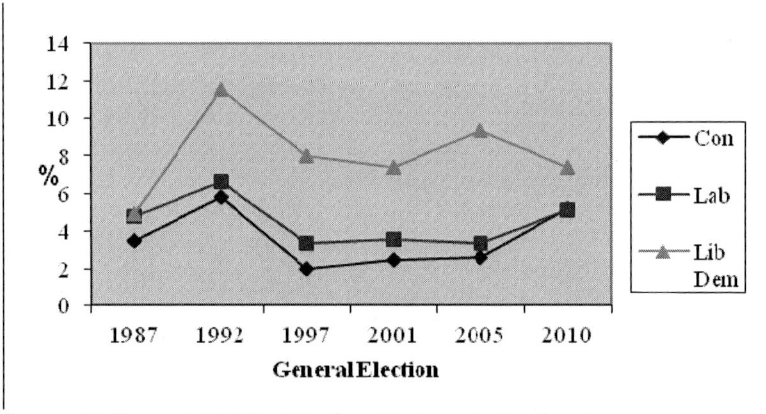

Source: Volkens *et al* 2012, http://manifesto project.wzb.eu/

[2] The Manifesto Project provides content analysis of national party manifestos in 55 countries, literally counting the number of sentences and coding them using a common scheme. The methodological approach assumes that issue salience can be measured by how often it is mentioned in the manifesto. If a party is treating an issue seriously, therefore, it should secure prominence in the election manifesto.

From 1992 the Liberal Democrats broke ranks by trying to establish the party as the greenest of the three major parties by adopting more progressive policies and making the environment a priority campaigning issue in successive elections, as is evident in Figure 1. Environmental protection became a sufficiently strong and consistent feature of party programmes for it to be regarded as a key element of Liberal Democrat ideology and political identity (Webb 2000: 106; Russell and Fieldhouse 2005). However, in a political system still dominated by the Conservative and Labour parties the Liberal Democrats struggled to impose their preferences on the political agenda. Even during election campaigns, despite prioritising the environment in leaders' speeches, press conferences and publications, the Liberal Democrats were unable to turn it into an issue of party competition. Indeed, polls consistently showed that the Liberal Democrats were unable to convince the public that they were the clearly the "greenest" of the three major parties, with Labour just as likely to come out ahead (Carter 2006: 760).

The cyclical mid-term character of environmental politics was less apparent during the first two Labour Governments (1997-2005) because there were none of the flurries of interest that had characterised earlier cycles. Indeed, Labour showed most interest in the environment in the weeks immediately following its electoral landslide. Promising to be the "greenest government yet", Tony Blair and John Prescott played prominent roles during the negotiation of the Kyoto Protocol at the COP-3 climate conference, but the Government soon lost interest in the environment. Labour's domestic environmental record during the first nine years in office was characterised by continuity rather than change. Labour was responsible for some important initiatives, including a Climate Change Levy (2001) imposed on non-domestic users and a Renewables Obligation (2002) to encourage electricity generation from renewable sources, but there was some complacency about its approach to domestic climate and energy policy (Carter and Ockwell 2007).

Yet, despite the low priority of the environment for the Government, which potentially made it vulnerable to political attack on this issue, the Conservatives gave Labour an easy ride. Under the leadership of William Hague (1997-2001) the Conservatives displayed little concern about environmental problems. On the contrary, an increasingly Eurosceptic party was often critical of progressive environmental legislation originating in Brussels. During the fuel protests that almost brought the country to a halt in September 2000 Hague praised the farmers and the hauliers leading the action as "fine upstanding citizens" (*The Guardian*, 18 September 2000). Hague's opening salvo in the 2001 election campaign was to promise a sixpence per litre cut in fuel duty, while his shadow chancellor, Michael Portillo, pledged to abolish the Climate Change Levy. During the brief sojourn of Iain Duncan Smith (2001-2003) the party leader did engage idiosyncratically with a limited range of environmental concerns, including solar energy, but without any lasting impact. Under Michael Howard (2003-2005) the environment again almost disappeared from view.

Explaining the limited party politicisation of the environment pre-2006

Four broad factors explain this limited party politicisation of the environment: political salience, interests, institutional and ideology.

First, if we assume that the self-interest of parties is to seek office then the bottom line is that low *political salience* of the environment amongst the public has dissuaded parties from an enthusiastic embrace of environmentalism. Whilst British people care about the environment, few people regard it as one of the most important issues facing the country and it has never been a significant determinant of the way people vote in general elections. Figure 2 shows that during Labour's second term in office (2001-2005) the share of people regarding the environment as one of the most important issues facing Britain generally stayed between just 2%-4%, rising slightly to hover around 6% during most of 2005. The focus of the electorate remained firmly on traditional materialist concerns such as the economy, taxation, health care, education and welfare policy. Consequently, government and opposition parties alike (in the UK and elsewhere) have been reluctant to champion environmental policies such as punitive regulations or green taxes that will impose short-term costs on companies or individuals in return for some imprecise, invisible long-term benefit. In short, British politicians, like their counterparts elsewhere, will normally focus on tomorrow's newspaper headlines, next week's polls and any upcoming election, rather than greenhouse emission levels in 2020, let alone 2050.

Figure 2. Percentage of people identifying the environment as a major problem

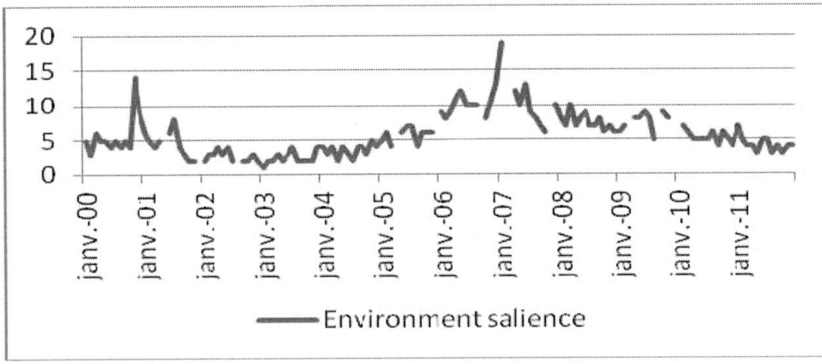

Source: Ipsos-Mori
http://www.ipsosmori.com/researchspecialisms/socialresearch/specareas/politics/trends.aspx

Second, a key *institutional* factor, the plurality electoral system, has largely insulated the major British parties from serious partisan threat from environmentalism. The bias of the electoral system against small parties with a geographically dispersed constituency of supporters has made it extremely hard for the Green Party to flourish. During the 1980s and 1990s the Green Party attracted less than 1.5% of the vote share in those seats it contested, failing everywhere to reach the 5% threshold needed to save its £500 deposit. However, after Labour entered Government in 1997 the Green Party slowly strengthened, building a small base of local councillors, often concentrated in university towns such as Brighton, Lancaster, Norwich and Oxford, which started to pay dividends in general elections. In 2001 the Greens secured a 2.9% vote share in seats contested and saved 10 deposits, rising to 3.4% in 2005 when 24 deposits were saved (Carter 2008; Birch 2009). The profile of the Greens also benefited from the constitutional reforms of the first Blair administration that introduced proportional representation across a range of second order elections resulting in the Greens securing regular representation in the European Parliament and the newly formed Scottish Parliament and Greater London Assembly from the late 1990s. Despite this boost, the Green Party remained a marginal force in electoral politics.

Environmental politics in Britain, by contrast with many European countries, was therefore largely non-partisan with very little party competition around the issue. Thus Labour and Conservative party strategists needed to focus attention on the environment only when there was an upsurge in public concern, confident that their main political opponent would be unable to gain political capital on the issue.

Third, this resistance to party competition was reinforced by pressure groups representing the economic *interests* of capital and labour. Business and trade union interests have frequently expressed their (often united) opposition to proposed environmental measures that will have a detrimental impact on the competitiveness of British industry, thereby threatening profits and jobs. One example was the concerted and robust opposition of many industries when Chancellor Gordon Brown first proposed his Climate Change Levy in 1999, which proved a bruising experience. Followed swiftly by the fuel protests led by farmers and hauliers in autumn 2000, a chastened Brown was put off introducing further eco-taxes for several years.

Although Britain has a large professional environmental lobby, its political influence during the first two terms of the Labour Government was quite limited compared to the access available to major business actors. Between 2000 and 2005 relations between the Chancellor and the green lobby were particularly frosty because Brown felt he had not received the public support he deserved from the environmental non-governmental organisations (ENGOs) over the Climate Change Levy (which was criticised by groups like Friends of the Earth for being too weak).

There are also *ideological* reasons why both the Conservative and Labour parties have found it difficult to embrace the environment wholeheartedly. Historically, the close links of the Conservatives and Labour with, respectively, business and the trade unions, made them unreceptive to environmental concerns that threaten their productivist pro-growth priorities of these interests. Moreover, the neoliberal ideology of the post-Thatcherite Conservative Party appeared unsympathetic to a progressive environmental agenda of regulations and green taxes. Whilst old left parties are generally regarded as more open to green issues than centre-right parties (Rohrschneider 1993), unlike its sister Social Democrat parties elsewhere in Europe New Labour enthusiastically embraced many aspects of neoliberalism whilst being "fundamentally suspicious of environmentalism" for being anti-aspirational, anti-consumerist and anti-growth (Jacobs 1999: 9). By contrast, the Liberal Democrats have never had these same productivist links with capital and labour, which helps explains their willingness to embrace the issue and make it a core part of their ideology.

Together these factors explain the limited party political interest in the environment and climate change before 2006: so why did it then change so dramatically?

The transformation of climate politics

From around 2006 changes in these factors combined to produce a fundamental transformation in the politics of climate change, which saw unprecedented party competition on this issue.

First, the bottom-line moved: the *political salience* of the environment and, specifically, climate change, grew rapidly, as indicated by increased media coverage, ENGO activity and public concern. As Figure 3 shows, press coverage, measured by headline stories related to climate change, increased significantly during 2005 and 2006, albeit still very small by comparison with traditional core issues such as "health" (Gavin 2009: 768-9). One stimulus for this increased attention was the stream of scientific reports that mostly presented the issue as cause of major concern. The media coverage intensified in late 2006 with the publication of the Stern Report (Stern 2007) in October followed by the release of the Fourth Intergovernmental Panel on Climate Change (IPCC) reports in February 2007 (IPCC 2007). The IPCC represented the epistemic community of concerned climate scientists by ensuring that the issue was picked up by the global media.[3] In the UK climate change stories were highlighted as deliberate editorial policy by the liberal media, especially the BBC, *The Guardian*, *Financial Times*, *Sunday Times* and *The Independent* (Ereaut and Segnit 2006). As the quantity of media coverage grew a consensus emerged in the public arena that climate change exists, that humans have contributed to it and that humanity will be harmed by it: a new common sense understanding seemed to have emerged.

---

[3] After the "Climategate" controversy (when the emails of University of East Anglia scientists were hacked into in a [failed] attempt to reveal a scientific conspiracy apparently faking the evidence of climate change) and the identification of some inaccuracies in the IPCC reports in 2009, UK media reporting of climate change stories became more critical and often sceptical.

Figure 3. UK Newspaper coverage of climate change or global warming 2000-2011

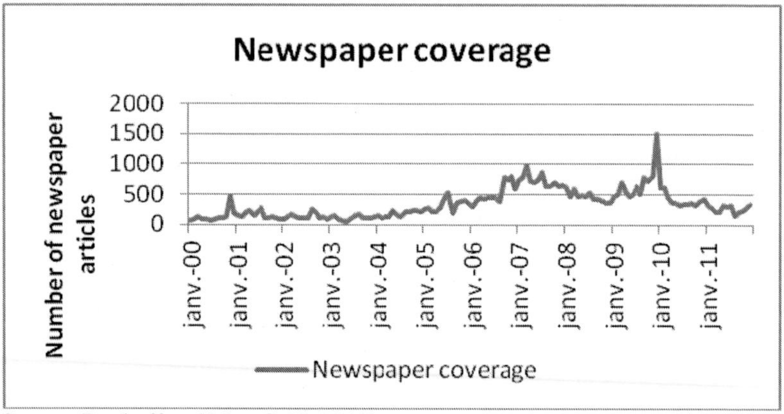

Source: Boykoff and Mansfield (2012)
http://sciencepolicy.colorado.edu/media_coverage/uk/

The media's growing appetite for climate stories was fed by very effective ENGO campaigning and lobbying that increasingly focused on climate change above other environmental issues. In particular, through the launch of its "Big Ask" campaign in May 2005 Friends of the Earth (FoE) alighted upon a simple policy discourse and associated demand: annual carbon targets and a Climate Change Bill. The "Big Ask" turned previously complex climate policy into a simple and focused campaign that began to mobilise large numbers of supporters. But its main objective was to win cross-party support. FoE recruited six MPs representing the three major parties and Plaid Cymru (including John Gummer, former Conservative Secretary of State for the Environment, and Michael Meacher, former Labour Minister for the Environment) to put down an Early Day Motion (EDM) in parliament calling for a bill to make the emissions reduction targets statutory (EDM 178, tabled 24 May 2005). Wider efforts at mass mobilisation included the formation in July 2005 of the Stop Climate Chaos coalition of NGOs, which brought together over 100 environmental, transport, development, faith and women's groups, and several major trade unions, to campaign for policies aimed at limiting temperature increases to no more than 2°C. But, as we shall see, it was the "Big Ask" that proved critical in shaping change.

The media coverage and ENGO campaigning together contributed to growing levels of public concern about climate change. One autumn 2005 poll found that 62% of people thought that every possible action should be taken to limit climate change and a further 32% believed some action should be taken (Ipsos-Mori 2006). Regular DFT surveys of attitudes to climate change and transport show that the proportion of people at least "fairly concerned" about climate change rose steadily to peak at 81% in 2006 of whom 87% were willing to change their behaviour to help limit it. This concern was manifested in the growing political salience of environmental issues generally. Most polls, especially those with time series data, conflate climate change with wider environmental issues. Thus regular Ipsos MORI polling (see Figure 2) shows that political concern about the environment started edging up during 2005 before rising rapidly in 2006-2007, peaking at 19% in January 2007 (making it the sixth most important issue facing the country) soon after the November 2006 spike in media coverage of the Stern Report and in the run-up to the publication of the Fourth IPCC report (Gavin 2009: 769). Although the levels of concern did decline they remained reasonably buoyant throughout the period of Labour Government.

Another important development was the discernible shift in attitude amongst producer *interests*, notably the business community, which began to see action on climate change as compatible with its commercial interests. One reason for this change reflected a willingness by some business leaders to embrace ecological modernisation: they saw that energy efficiency cuts costs and increases productivity; that emissions trading with free allowances can create valuable marketable assets; and that investment in green technologies could give them a competitive edge in existing markets or create new markets for their products. Many other businesses simply saw that a positive and progressive approach to climate policy was good for public relations and not especially costly. Of course, many businesses continued to oppose action because of the perceived costs of climate mitigation policies, but a significant group of businesses "switched sides", preferring to influence (and perhaps moderate) policy from within the supporters' camp rather than be seen as oppositional and thereby lose influence with Government (especially as they noted that the Opposition leader, David Cameron, had embraced climate change too – see below).

This shift in attitude was expressed in several ways. The formation of the Corporate Leaders Group in 2005 showed the engagement of leaders of some of the largest and most powerful corporations. An open letter from this group to the Prime Minister in June 2006 calling for tougher carbon reduction targets to be set under the European Union (EU) Emissions Trading Scheme as a means of stimulating investment in low carbon technologies was an important statement of intent (Corporate

Leaders Group 2006). Following the publication of the Stern Report in November 2006 several companies announced new carbon reduction initiatives, such as the Plan A commitment by Marks and Spencer to be carbon neutral by 2012. Another Corporate Leaders Group public letter addressed to the three main party leaders in September 2008 urged them to reach "a cross-political consensus on the scale and speed of change required" to deliver a "transformational" approach to climate change, and not to be dissuaded from taking action by the financial crisis (*The Guardian,* 19 September 2008). Also significant, because it better represented the wider business community, the Confederation of British Industry, the major UK employers' organisation, published a landmark report supporting action on climate change (CBI 2007). These developments coincided with the growing recognition that new infrastructure investment was needed in electricity generation, so energy companies realised that they could make money out of whatever technologies they were mandated to use, as long as the regulatory framework was clear and certain. Thus, with the business community considerably less hostile, there was much greater scope for the major parties to advocate tougher climate measures.

Meanwhile, the trade union movement had also become much more engaged with environmental issues and climate change, notably through its green workplace strategy to encourage greater employee participation in developing sustainable workplaces (Farnhill 2013). The consensus across both sides of industry was illustrated by the enthusiasm with which the General Secretary of the Trades Union Congress (TUC), Brendan Barber, welcomed the publication of the CBI report "We are convinced that there is a green industrial strategy to be won from an effective response to the huge challenges of climate change. The CBI is right to point to the need for action on energy efficiency at work." (TUC 2007).

Thus changes in a range of factors exogenous to the party system helped create the conditions for party competition on climate change. But the major parties themselves played a vital role in overcoming some of the obstacles.

Arguably, the catalyst for change was David Cameron's decision to make the environment the signature issue in his programme for the modernisation of the Conservative Party. After becoming party leader in December 2005, Cameron immediately adopted a political marketing strategy, masterminded by his political advisor, Steve Hilton, to "detoxify" the Tory brand. The environment was identified as a key component in this decontamination strategy that aimed to remove the "nasty party" image by emphasising issues, such as the environment, that Conservatives had previously tended not to talk about, and which might appeal to women, younger voters and Liberal Democrats (Carter 2009; Bale 2010). The environment was to be used for "repositioning the Conservative Party in the public's imagination, about showing it was changing" (Bale 2010, p.290). A series of well-publicised media events ensued, including a trip on an Arctic dog sled to observe the effects of global warming, cycling to work and an application for planning permission to place a wind turbine on his house. The Conservative Party's red, white and blue "torch of freedom" logo was replaced by an oak tree with green foliage and a blue trunk. These symbolic gestures were backed up by a sustained rhetorical commitment in keynote speeches and the decision to fight repeated local election campaigns under the slogan "Vote Blue, Go Green". As a political marketing strategy this embrace of the environment certainly seemed to resonate with the political mood of the time. More debatable is the extent to which it represented a genuine attempt by a small coterie of leading Conservatives to alter the ideology of the party.

Yet the Labour Government had never entirely ignored the environment. Indeed, a strange dichotomy characterised its approach to climate change during 2004-2006. On the one hand, it was widely acknowledged that no world leader had done more to push climate change up the global agenda than Tony Blair, notably by his decision to make it a priority at the G8 summit in Gleneagles in July 2005 and during the UK Presidency of the European Union. In preparation for the G8 Blair lectured business leaders in a speech in September 2004 on the scientific and economic case for taking action on climate change and announced that the Government would sponsor a major conference on the science of climate change at the Meteorological Office in Exeter in February 2005. Yet, on the other hand, Britain's domestic record on climate change came under increasing critical scrutiny during the same period (Carter and Ockwell 2007).

Just as Conservative polling encouraged Cameron to embrace the environment, Labour strategists were increasingly aware there were good electoral reasons for Labour to strengthen its green credentials. The growing political salience of environmental issues was also expressed in an increasingly buoyant Green Party. The profile of the party had benefited from having a small group of elected Green representatives, including MEPs. The majority of deposits saved in the 2005 general election were in Labour-held seats, indicating that it was drawing support from disillusioned Labour voters, probably encouraged by the party's firm opposition to the Labour Government's involvement in the Iraq War (Carter 2008). By 2007-2008 there was evidence of a small but growing base in voter attitudes for more radical green policies. Private Labour polling showed that there was a group of voters for whom a party's green credentials were a partial determinant of their voting intentions; in short, green issues were electorally salient (Labour Party representative, personal interview 2010). Crucially, there were around 20 constituencies where the potential Green Party vote was comparable to the majority of the sitting Labour MP (including Brighton, Cambridge, Oxford, Islington South). In addition, extensive polling data showed that green concerns were present in sizeable numbers of other voters, notably young mothers and professionals.

Against this background, and with the Liberal Democrats keen to remind everyone of their longstanding commitment to the environment, the period 2006-2009 was characterised by unprecedented levels of party competition over climate change and the environment. As the parties developed their positions they started to outbid each other to be "greener" and it seems reasonable to claim that this intensification of party competition contributed to the transformation of climate and energy policy during this same period. The extent of party politicisation of climate change, and also its impact on policy outputs, can be illustrated by three high-profile issues.

First, the Government announced a Climate Change Bill in the Queen's Speech on 15 November 2006. When David Miliband became Secretary of State for the Department of Environment, Food and Rural Affairs (DEFRA) in May 2006 the Labour Government was under growing pressure to strengthen its climate and energy policy, but he had no specific plan to introduce a climate change bill. However, on 1 September 2006 FoE announced a major political coup: it had persuaded David Cameron to turn his green rhetoric into policy substance by signing up to the Big Ask. By the end of the month FoE had persuaded 412 MPs to sign its Early Day Motion (a parliamentary mechanism used by MPs to draw attention to an issue or event), the fifth highest total for an EDM since 1939 (House of Commons 2010). With the Liberal Democrats already supporting it, Cameron's shift placed huge pressure on the Labour Government to respond. Within weeks David Miliband announced he would be introducing a Climate Change Bill. David Miliband's private comment that "We cannot be seen to be the only party not supporting this idea" (Labour Special Advisor, personal interview) provided a clear indication that for the first time the three major parties had started competing seriously to be greener than each other (*The Guardian* 26 October 2006).

Cameron maintained the pressure on the Government by taking strong positions on two more high profile climate related issues: the proposal to construct a third runway at Heathrow airport and the application by the multinational energy utility E.ON to build a new coal-fired power station at Kingsnorth, Kent.

The Heathrow runway was strongly opposed by local residents, ENGOs and the Liberal Democrats. In June 2008 Cameron declared that the Conservatives would oppose the proposal, a decision undoubtedly influenced by the presence of several Conservative-Liberal Democrat marginal constituencies in the area including Richmond Park where millionaire Zac Goldsmith, former editor of the Ecologist, was the prospective Conservative candidate. Although the Labour Government, strongly backed by the business community, eventually approved the new runway, the potential damage to Labour's climate reputation encouraged a fiercely fought rearguard campaign in Cabinet led by the Secretary of State for the new Department of Energy and Climate Change (DECC), Ed Miliband, and Hilary Benn, Secretary of State at DEFRA. They managed to win vital "environmental" concessions regarding aviation emission levels and the allocation of landing slots to low emission aircraft (Hasan and Macintyre 2011: 146-9). The Conservative-led Coalition Government subsequently overturned its predecessor's decision to approve the runway.

If approved Kingsnorth would be Britain's first new coal-fired power station in three decades. A vociferous ENGO campaign against the plant was backed by the Conservatives and Liberal Democrats. Initially, the Business Secretary, John Hutton, was minded to approve E.ON's application, but when the energy portfolio was integrated into DECC under Ed Miliband in October 2008, and the political pressure on the Government was stoked up by both main opposition parties, Ed Miliband required a solution compatible with the Government's climate commitments. In April 2009 he announced that no new coal-fired power stations would be permitted unless they incorporated the new technology of carbon capture and storage (CCS). The Government would finance up to four CCS demonstration projects, funded by a levy amounting to a 2% increase in consumer energy bills by 2020.

Alongside these high profile controversies for a brief period the three major parties were competing with each to offer more progressive climate mitigation policies than their rivals. In particular, it was a rare moment in ENGO history because they almost had *carte blanche* to persuade the Conservatives to adopt pro-climate policies as a way of pressurising Labour. Several of the Government's green initiatives were policies initially proposed by the Conservatives, including the Climate Change Bill, reforms of air passenger duty, feed-in tariffs and support for a new high-speed railway linking London to Birmingham and beyond, suggesting that their adoption by the Labour Government was at least partly shaped by party competition.

From "bad politics" to "good politics"?

The factors that had dissuaded the major British parties from promoting progressive climate policies weakened or disappeared during 2006. Suddenly the issue had political salience: the public, the media, the ENGOs and even the business community demanded action. No longer was climate politics "bad politics" whereby even mild policy initiatives were met by fierce criticism from all sides. Now the parties were scrambling to catch up with the political mood and began competing aggressively to prove they had better green credentials than their opponents. Indeed, by 2008-2009 the Labour and Conservative parties had shifted from their traditional strategy of preference accommodation to a preference-shaping strategy. Climate politics became increasingly elite driven, with ministers such as David and Ed Miliband, and David Cameron, arguably now ahead of public opinion as illustrated by their support for a massive expansion of onshore and offshore wind energy, the construction of a new generation of nuclear power stations and acceptance that such policies would significantly increase domestic energy bills.

This party politicisation undoubtedly contributed to the step-change in climate and energy policy that started in mid-2006. Of course, many other factors combined to persuade the Government to carry out its fundamental transformation of climate and energy policies, including the strengthening scientific evidence, the Stern Report, tough EU emission reduction targets, the 2009 Copenhagen COP-15 climate summit and pressure from public concern, ENGOs and sections of the business community. But one consistent theme in all of the elite interviews was the significance of the "Cameron effect" on Government policy. His support for stringent climate measures placed enormous pressure on the Government to respond, but by the same score Cameron's support made tough decisions (like introducing a Climate Change Bill or placing a levy on energy bills to fund CCS) easier for David and Ed Miliband to sell to their sceptical Cabinet colleagues by removing the risk of Opposition party attack. Indeed, as senior ministers gave increasing priority to the problem, climate change became part of the ideological commitment of the Labour Government, not just a peripheral policy area. This shift was even more apparent after the financial crash, when the rhetoric of the low carbon economy was given substance in the form of a major green stimulus package, providing public finance that might not otherwise have been forthcoming, authorised by Labour ministers not known for supporting green causes, such Peter Mandelson and Alistair Darling.

The intense party competition over climate change dissipated in the run-up to the 2010 general election, in the wake of Climategate and the disappointing COP-15 climate summit at Copenhagen. The issue barely figured during the election campaign or the first thirty months of Coalition Government. Moreover, the cross-party consensus in favour of radical action on climate change seemed to have fractured, as opposition to costly emission reduction measures and 'unsightly' wind farms spread from the Conservative backbenches up to the Chancellor George Osborne. If the party politicisation of climate change proves to be a temporary blip, at least it contributed to the enduring institutional legacy of this period provided by the Climate Change Act, which has (so far) ensured that there remains considerable momentum behind UK climate policy.

Bibliography

Bale, Tim, *The Conservative Party from Thatcher to Cameron*, Cambridge: Polity, 2010.

Birch, Sarah, 'Real Progress: Prospects for Green Party Support in Britain', *Parliamentary Affairs*, 62(1): 53-71, 2009.

Boykoff, Max & Mansfield, Maria, *2000-2011 UK Newspaper Coverage of Climate Change or Global Warming*, Boulder: University of Colorado, Center for Science and Technology Policy Research, http://sciencepolicy.colorado.edu/media_coverage/uk/, 2012.

Carter, Neil, 'Party Politicisation of the Environment in Britain', *Party Politics,* 12(6): 747-67, 2006.

—. 'The Green Party: Emerging from the Political Wilderness?' *British Politics*, 3: 223-240, 2008.

—. 'Vote Blue, Go Green? Cameron's Conservatives and the Environment', *Political Quarterly*, 80(2): 233-242, 2009.

Carter, Neil & Ockwell, David, *New Labour, New Environment?* York: Centre for Ecology, Law and Policy, http://celp.org.uk/projects/foe/, 2007.

Compston, Hugh & Bailey, Ian *Climate Clever: How Governments Can Tackle Climate Change (and Still Win Elections).* London: Routledge, 2012.

Confederation of British Industry. *Climate Change: Everyone's Business.* London, 2007.

Corporate Leaders Group on Climate Change, Letter to the Prime Minister, http://www.cpsl.cam.ac.uk/Business-Platforms/The-Prince-of-Wales-Corporate-Leaders-Group-on-Climate-Change/UK-CLG.aspx, 6 June 2006.

Ereaut, Gill & Segnit, Nat, *Warm words: how are we telling the climate story and can we tell it better?* London: Institute for Public Policy Research, 2006.

Farnhill, Thomas, 'Green No More? The Coming-of-Age of UK Trade Unions' Environmental Agenda, 1970-2011', PhD thesis, University of York, 2011.

Gavin, Neil, 'Addressing Climate Change: A Media Perspective', *Environmental Politics*, 18: 765-80, 2009.

Hasan, Mehdi & Macintyre, James, *Ed: the Milibands and the making of a Labour Leader*, London: Biteback, 2011.

House of Commons. 2010. *Early Day Motions.* Factsheet. http://www.parliament.uk/about/how/guides/factsheets/procedure/p03/

Intergovernmental Panel on Climate Change, *Working Group III contribution to the Intergovernmental Panel on Climate Change Fourth Assessment Report. Climate Change 2007: Mitigation of Climate Change. Summary for Policymakers,* Geneva, 2007.

Ipsos-Mori, *Political and Social Trends,* http://www.ipsos-mori.com/researchspecialisms/socialresearch/specareas/politics/trends.aspx, 2013.

Jacobs, Michael, *Environmental Modernisation: The New Labour Agenda*, London: Fabian Society, 1999.

Rohrscheider, Robert, 'New Party Versus Old Party Realignments: Environmental Attitudes, Party Policies and Partisan Affiliations in Four West European Countries', *Journal of Politics.* 55(3): 682–701, 1993.

Stern, Nicholas, *The Economics of Climate Change*, Cambridge: Cambridge University Press, 2007.

Russell, Andrew & Fieldhouse, Edward, *Neither Left Nor Right? The Liberal Democrats and the Electorate*, Manchester: Manchester University Press, 2005.

TUC, *TUC comment on CBI green report,* http://www.tuc.org.uk/social/tuc-14015-f0.cfm, 26 November 2007.

Volkens, Andrea *et al*, *The Manifesto Data Collection*. Manifesto Project (MRG/CMP/MARPOR), Berlin: Wissenschaftszentrum Berlin für Sozialforschung, http://manifesto-project.wzb.eu/, 2012.

Weale, Albert *et al*, *Environmental Governance in Europe*. Oxford: Oxford University Press, 2000.

Webb, Paul, *The Modern British Party System*, London: Sage, 2000.

ENVIRONMENTAL ISSUES IN POLITICAL SPEECH: DISCOURSE ANALYSIS OF THREE BRITISH PRIME MINISTERS

HÉLÈNE LEDOUBLE AND OLIVIER GOUIRAND
UNIVERSITY OF TOULON, BABEL RESEARCH LABORATORY

This article focuses on transcripts of speeches pronounced by the last three British Prime Ministers, from 1997 to 2011. In this particular study, our objective is to identify each PM's positioning regarding environmental issues through the linguistic analysis of their discourse.

Each PM leaves a mark on History. The content of their discourse is the reflection of society at a given time and their vision of that society. Words and concepts involved may convey the different perceptions each PM has on some issues or their will for action.

In the first part, this paper will reveal the results of the lexicometric analysis of the corpus and the related ontological content. This lexical and conceptual approach will highlight the salient features of these political texts.

In a second part, we will analyze the relation between these speeches and the way environmental issues are tackled: a global approach to the different types of discourse and argumentation will precede a more lexical approach related to the choice of words and their specific meanings.

We will eventually question the major differences between speeches. Can we talk of temporal or cyclical distinctions between them?? Are they more political differences between the Labour and Conservative Party, or simply linguistic or diachronic variations? This article will attempt to answer these questions before delivering more general perspectives to this analysis.

Corpus

Our corpus comprises speeches by three British Prime Ministers during their political mandate:

- The corpus related to Tony Blair includes speeches pronounced between 1997 and 2007. It was extracted from an archive of the PM official website http://www.number10.gov.uk/. It has 627,404 words.
- The corpus related to Gordon Brown includes speeches pronounced between 2008 and 2010. It was extracted from his book "The change we choose" and comprises 124,107 words.
- The corpus related to David Cameron includes speeches pronounced between April 2010 and December 2011. It was extracted from the PM official website http://www.number10.gov.uk/. It has 416,372 words

In order to moderate differences between corpus sizes, all the statistical analyses will be based on relative rather than absolute frequencies.

General themes in the 3 types of discourse

To get an overview of the general themes present in the three corpora, we used the scenario tool of the Tropcs software[1]. The general thesaurus used for corpus analysis shows the major differences between the three types of discourse as presented in the graph below[2] (the color blue is used for Cameron, orange for Brown and green for Blair).

The prominent spikes in blue represent the recurrent "Business & Commerce" issues as well as "Money, Taxes and Expense" being tackled by David Cameron. For Tony Blair, "Education" is prominent, as well as "Law and Justice".

For Brown, the spike of the "Crisis & disorganization" concept is blatant, due to the raging international economic crisis in the years 2010-2011. The "Energy" and "Weapon" concepts are also pervasive, Brown being extremely concerned by the nuclear issue as we shall develop later.

The "Employment and work" concept is almost equally important in the three corpora.

---

[1] Molette, Pierre and Landré, Agnés, http://www.semantic-knowledge.com, version 8.0, 2011.

[2] Graph created with MS Excel using data from Tropes.

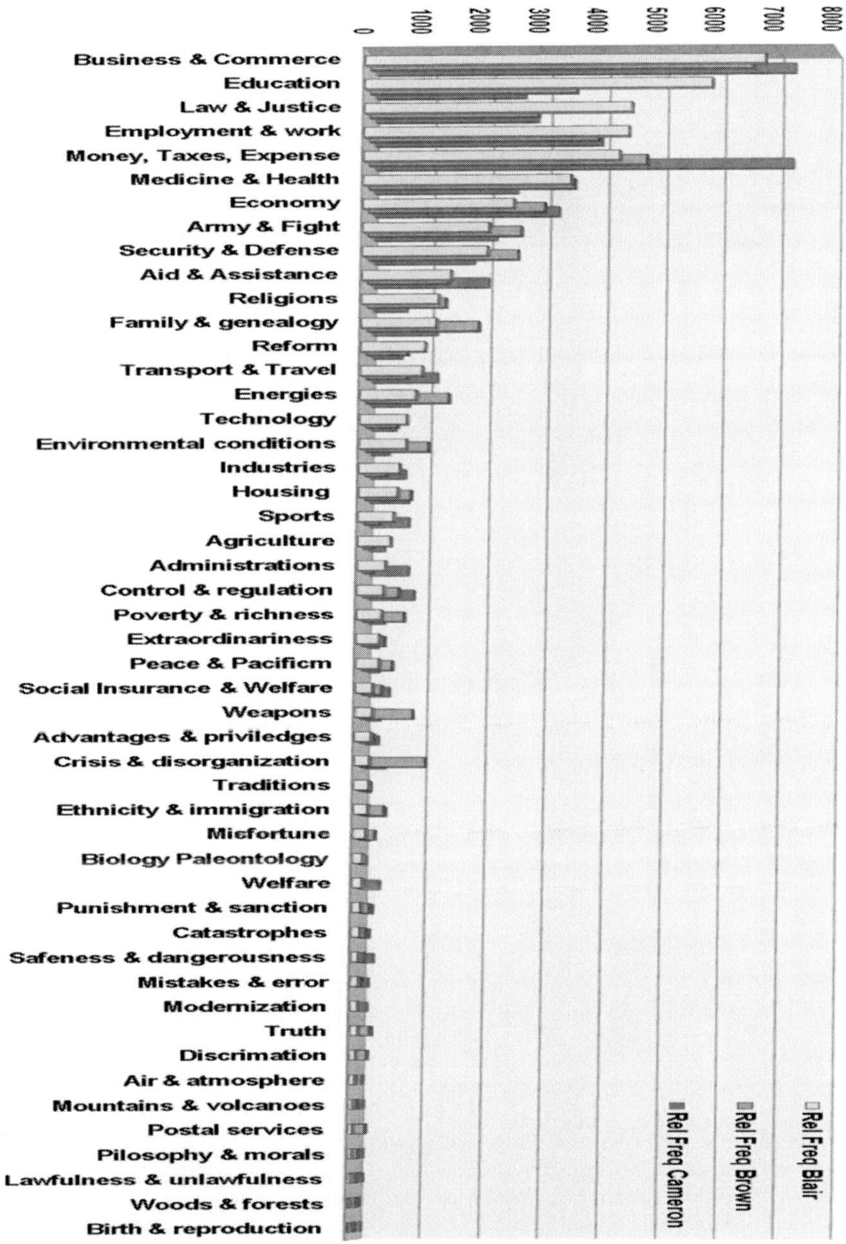
0
1000
2000
3000
4000
5000
6000
7000
8000
Business & Commerce
Education
Law & Justice
Employment & work
Money, Taxes, Expense
Medicine & Health
Economy
Army & Fight
Security & Defense
Aid & Assistance
Religions
Family & genealogy
Reform
Transport & Travel
Energies
Technology
Environmental conditions
Industries
Housing
Sports
Agriculture
Administrations
Control & regulation
Poverty & richness
Extraordinariness
Peace & Pacificm
Social Insurance & Welfare
Weapons
Advantages & priviledges
Crisis & disorganization
Traditions
Ethnicity & immigration
Micfortune
Biology Paleontology
Welfare
Punishment & sanction
Catastrophes
Safeness & dangerousness
Mistakes & error
Modernization
Truth
Discrimation
Air & atmosphere
Mountains & volcanoes
Postal services
Pilosophy & morals
Lawfulness & unlawfulness
Woods & forests
Birth & reproduction
Rel Freq Blair
Rel Freq Brown
Rel Freq Cameron

Fig. 1

Environmental themes in the 3 types of discourse

The three corpora were analyzed as follows: environmental themes were identified manually using the terms from the corpora sorted by frequency in the concordance program Textstat[3]. More specifically, the matches between terms and themes were established by checking the terms in the corpora against a cross-section of authoritative glossaries on environmental terms[4] which were then verified in context[5]. These terms were finally grouped into thematic categories, as presented in the table below:

[3] By Hüning, Matthias, TechnischeUniversität Berlin, version 2.8g, 2009.
[4] As part of our broader research project, a set of glossaries issued by international or governmental bodies, universities and organizations specializing in the Environment were compiled, similarities and differences were output.
[5] See Ledouble, Hélène and Gouirand, Olivier, "Environmental issues in news and scientific discourse. Diachronic approach to the notion of hybridity", Conference on *Green Growth*, University of Lyon, September 2011.

AIR	Air		49.41	0.00	0.00
	Atmosphere		14.34	0.00	0.00
	CO2 – carbon dioxide		36.66	0.00	0.00
	Greenhouse		117.95	0.00	2.40
	Smog		9.56	0.00	0.00
	Emission		304.43	185.32	24.02
	Total AIR		532.35	185.32	26.42
CLIMATE CHANGE-WEATHEF	(global) warming		47.82	32.23	4.80
	Climate (change)		479.75	692.95	103.27
	temperature		30.28	24.17	0.00
	Total CLIMATE		557.85	749.35	108.08
EARTH	earth		17.53	24.17	0.00
	soil		9.56	0.00	0.00
	planet		25.50	153.09	12.01
	Drought		9.56	32.23	7.21
	Total EARTH		62.16	209.50	19.21
ENERGY SOURCE	Fossil	fossil	15.94	0.00	0.00
		coal	19.13	24.17	9.61
		oil	31.88	48.35	12.01
	Ss-Tot		66.94	72.52	21.62
	renewable	renewable	82.88	56.40	28.82
		solar	17.53	16.12	7.21
		wind	15.94	0.00	16.81
		tidal	14.34	0.00	4.80
	Ss-Tot		130.70	72.52	57.64
	Nuclear	nuclear	51.00	322.30	55.24
		uranium	1.59	16.12	0.00
	Ss-Tot		52.60	338.42	55.24
	Total ENERGY SOURCE		250.24	483.45	134.50
TRANSPORT	transport		39.85	0.00	4.80
	congestion		9.56	0.00	0.00
	traffic		9.56	0.00	0.00
	road		4.78	0.00	0.00
	car		47.82	24.17	16.81
	bus		6.38	0.00	0.00
	vehicle		46.22	8.06	14.41
	railway		0.00	0.00	0.00
	trains		1.59	0.00	0.00
	Total TRANSPORT		165.76	32.23	36.03

Fig. 2

WATER	water	65.35	16.12	9.61
	sea	23.91	8.06	26.42
	Flood/ing	36.66	24.17	9.61
	Tsunami	9.56	0.00	0.00
	(Oil) spill	0.00	0.00	7.21
	Total WATER	135.48	48.35	52.84
BIODIVERSITY	biodiversity	19.13	0.00	0.00
	wildlife	14.34	0.00	0.00
	countryside	11.16	0.00	0.00
	forest	20.72	16.12	0.00
	species	9.56	0.00	0.00
	Deforestation	6.38	0.00	0.00
	Total BIODIVERSITY	81.29	16.12	0.00
POLLUTION	polluti/on/ng	0.00	0.00	0.00
	litter	19.13	0.00	0.00
	waste	30.28	16.12	2.40
	Total POLLUTION	49.41	16.12	2.40
ENERGY TYPE	electricity	19.13	0.00	12.01
	hydrogen	12.75	0.00	0.00
	fuel	68.54	0.00	14.41
	Total ENERGY TYPE	100.41	0.00	26.42
CONFERENCES	kyoto	183.29	0.00	0.00
	copenhagen	0.00	249.78	4.80
	Total CONFERENCES	183.29	249.78	4.80
QUALITY	sustainable	207.20	249.78	127.29
	green	113.16	112.81	141.70
	clean	100.41	32.23	7.21
	unsustainable	19.13	40.29	48.03
	environmental/ly	247.05	112.81	28.82
	natural	36.66	8.06	4.80
	Total QUALITY	723.62	555.97	357.85
MEASURES	recycl-e-ed-ing	27.10	0.00	0.00
	reduction (emission)	54.19	56.40	4.80
	Total MEASURES	81.29	56.40	4.80

Fig. 3

Naturally, the above-mentioned categories are inevitably fragmentary, since PMs cannot be deemed to have covered all topics in the field of the Environment; there is also no general agreement among experts on such a notion.

However, in the narrower context of political speech in the UK over a period of some fifteen years, a number of conclusions can be drawn from this table. Indeed, it goes to show that there are significant quantitative variations between the three PMs over the 12 categories:

- Overall, Tony Blair makes more use of environmental terms with an average 243.6 per million relative frequency (roughly one in 4,000 words), compared to 216.9 for Gordon Brown and especially to David Cameron whose frequency is 64.5 (one in 15,000+ words).
- A more fine-grained analysis shows that Tony Blair uses a larger spectrum of terms: all categories are represented and they are more evenly distributed than his successors. Conversely, Gordon Brown relies heavily on *climate* (*change*) and *nuclear* (*power*) which alone account for 39% of the total while David Cameron has categories with zero or very few hits (BIODIVERSITY, which seems to have gone out of fashion totally; POLLUTION, CONFERENCES and MEASURE being negligible – roughly one in 200,000 words each).

The three highest ranking categories (respectively CLIMATE, QUALITY and ENERGY SOURCE) are also telling because they demonstrate a general tendency for simplification (climate change is only one environmental issue in the broader picture), for descriptiveness (identifying problems does not mean solving them – this is supported by weak data in MEASURES) and skewed perspective on energy sources (the bottom line is the geopolitical stance on fossil fuels and reliance on nuclear power as a substitute thereof).

However, identifying themes and extracting semantic categories alone is insufficient to grasp the argumentative and lexical features needed to fully understand the meaning of political texts and the intentions of their authors.

British political speech and the Environment

Discourse and Argumentation

In *The Language of Environment*, Myerson & Rydin established a distinction between 3 types of Environment-related discourse:

1. Accountable and strategic (scientific or expert reports, for ex. the Brundtland Report, IPCC[6] Reports)
2. Ironical and/or pessimistic on potential solutions (contradictory reports).
3. Alarming, using emotion-tainted speech and fear (narrative speech)

[6] International Panel on Climate Change

Flottum also examines the scientific type of argumentation in her detailed study of the IPCC report: such discourse calls for the description of the observed phenomena in terms of climate change and the predictable future situation based on these observations. She then underlines the rational approach to climate change: after a short- and middle-term perspective, the scientific author adopts a long-term perspective, giving evidence that these changes are irreversible (if nothing is done). Based upon these scientific reports, she underlines the fact that politicians need to adjust and to act.

The narrative type of discourse is more "emotional": it focuses on the forthcoming danger and the correlated emergency to act: without effective action, consequences can be dramatic and world-wide. To give evidence of the imminent disaster, narrative speech takes the observed facts to an extreme where they become fiction. For instance, the alarming and occasionally 'fictional' scenario visually depicted in Al Gore's *Inconvenient Truth* can also be put into words in political speech.

We will now analyze the argumentative and narrative form of each discourse in relation to these approaches, taking into account its specificities and global consistency.

Tony Blair

Tony Blair uses both a strategic type of discourse based on scientific evidence and a more narrative and alarming speech focused on facts and fiction.

Blair centers argumentation on the postulate that the emission of greenhouse gases is causing global warming. He first lists all the potential impacts of this phenomenon:

> "Many species of plant and wildlife are expected to become extinct"
>
> "The UK will have wetter winters, more summer droughts, extreme weather and heavy rainfall"

The intensity and gravity of this evolution is conveyed through compelling structures as well as "passionate" lexical items:

> "The consequences for our planet are literally disastrous"
>
> "Global warming [...] has become alarming and is simply unsustainable in the long-term"

Blair uses the dialectic image of a damaged earth that is already being destroyed by Man and will slowly disappear if nothing is done to protect it. Blair's strategy is to announce a "state of emergency". The more serious the situation is, the more imperious, pressing and undisputable the solution.

Another communicative strategy used by Blair is a repetitive pattern of structures conveying the dangerous absence of action and the induced dramatic consequences:

> "Without radical international measures to reduce carbon emissions within the next 10 to 15 years, there is compelling evidence to suggest we might lose the chance to control temperature rises"
>
> "Failure to act will make an increase of between 2 and 5 degrees in average temperatures almost inevitable"

The argumentation is complemented by repeated injunctions urging for action:

> "We must implement the energy review"
> "We must develop far cleaner and more efficient aircraft and cars"
> "It is now that timely action can avert disaster"

Blair also engages in proselytizing Britain's attitude:

> "Britain will never be a modern, forward-looking country if it is a place whose beauty, character, air, rivers, are polluted, defaced, and contaminated. To be modern is to be green"
>
> "We need to go far beyond Kyoto, in all probability reducing greenhouse gas emissions by 60% by the year 2050"

He fights the idea that a country's growth is opposed to sustainability:

> "And it is a myth that reducing emissions makes us poorer"
> "If you look in the 12 years 1990 to 2002, we in Britain cut our emissions by about 15 percent whilst we were growing at 30 percent"

He also discusses the "alleged fact" that sustainable development is only devoted to wealthy people and countries:

> "And it is through our actions we demolish the myth that sustainable development is only important for the well off–that have the time and resources to be green"

One characteristic of Blair's vision is a belief that action can be taken to face and solve the situation, without changing our current lifestyle:

> "Such action can be taken without disturbing the essence of our way of life, by adjusting behaviour not altering it entirely"
>
> "The world could reduce emissions by over 60%. This would not involve huge shifts in the economy or enormous changes in lifestyles"

Also, Tony Blair is concerned with the culprit behind climate change, that is, emissions. His statements reflect the idea that international cooperation can solve this problem, provided the "biggest economies" join their forces:

> "If Britain shut down our emissions entirely i.e. we closed down the country, the growth in China's emissions would make up the difference within just two years. Without the biggest economies being part of a framework to reduce carbon dependence, we have no earthly hope of success"

This highly debated approach, central to Environment-related questions has, at times, appeared to scientists as a utopian more than a rational picture of reality. This controversial optimism is however predominant in Blair's discourse.

Gordon Brown

The major environmental themes tackled by Brown are climate change and energy supply, as well as the concept of Earth/Planet and its protection.

1- Climate change and energy supply are viewed from a an economic standpoint *(energy source, climate finance)* or a security one (*energy security, nuclear energy) :*

> "As you know, I have put forward a package of proposals for how such climate finance could be organised, with a working figure of $100 billion per year in predictable public and private funding by 2020"
>
> "Iran has a clear choice to make suspend its nuclear programme and accept our offer of negotiations or face growing isolation and the collective response not of one nation but of many nations. And we will do more than oppose what is wrong. We will show those who would give licence to terror the way home to what is right"

Here are possible explanations for this. As the Chancellor of the Exchequer for 10 years in the Blair government, he came to be an expert in economic and financial matters. Besides this, his mandate corresponded to the beginning of the world economic crisis.

The reasons for the security issue are not as clear. One pervasive threat for European nations is the diplomatic difficulties and tense relations with Iran, which put pressure on energy supply and the cost of energy.

The approach to the question is geopolitical. For instance, Brown painstakingly tries to show the advantage Iranian citizens can get from a civil nuclear program controlled by Western countries which possess the needed expertise:

> "Iran therefore faces a clear choice to continue in this way and face further and tougher sanctions or change to a UN-overseen civil nuclear energy programme that will bring the greatest benefits to its citizens"

2- The Earth/Planet is prevalent in Brown's discourse. Fifty percent of occurrences are comprised in a *save/preserve the planet* collocation. In all other cases, the use of these terms is associated to anxiety-inducing situations (regular association to world issues such as Terrorism or nuclear proliferation).

But Brown generally remains quite vague. The facts and concrete measures are rarely detailed.

The corpus remains descriptive: the semantic category QUALITY (including general terms such as *environmental, green, sustainable,* etc.) stands out in this corpus. For instance, the use of adjectives along with the notions of aim (the operator *to*) and volition (the verb *want*) produces an incantatory message:

> "I want this to become the moment when together we rise to the new challenges by purposeful visionary and international leadership, leaving behind the orthodoxies of yesterday and embracing new ideas to create a better tomorrow not as victims of history but as shapers of an open, free-trade, flexible globalisation that is also inclusive and sustainable"

There is a hierarchical and simplistic approach (developed country versus poor country, underdeveloped nation versus resources-rich nation, Ethics versus Terrorism etc.)

Summits and Conferences are mentioned with their lot of "government" measures, which will be hard to impose in times of economic crisis, except for the creation of The Department of Energy and Climate Change (DECC) in 2008; therefore, they can be considered as an illusory way to solve problems.

> "That is why I have proposed a new approach to financing our action against climate change, which will provide substantially increased, additional and predictable flows, from both public and private sectors, of around $100 billion a year by 2020"

Brown's office is characterized by the pessimistic context of the financial crisis with no financial margins to carry out any major environmental program initiated by his predecessor or himself, thus creating a discrepancy between discourse and fact. Let us now see how his successor goes about the issue.

David Cameron

A general analysis of Cameron's type of discourse leads to 3 general observations:

1- He adopts a non-specialist positioning

As we can see with the examples below, Cameron is not as assertive as Tony Blair, with questions related to the Environment. Indeed, he makes an intensive use of

- *experts' opinions:*

> "I think if you add up all the things that the different countries have said they ought to do, if you add them all up, we would probably limit global warming to two degrees, which is what the experts say is dangerous but acceptable"

- *modals and signs of uncertainty*

Adverbs expressing uncertainty (probably, maybe, etc.) and modals are common, especially, regarding the new Department of Energy and Climate Change:

> "It's basically going to, I think, employ thousands of people up and down the country in important environmental work. It should save people money on their energy bills. It should cut carbon emissions. It should help reduce fuel poverty"
>
> "I think the new Department of Energy and Climate Change works"

- *simple lexicon and vague sentences*

When talking about environmental matters, Cameron does not seem to use many technical or specialized terms:

> "In terms of insulation and things like that, the Green Deal [...] is in all our interests"

He often mentions a personal "enthusiasm" rather than scientific arguments:

> "This is the Green Deal that we're-establishing. [...] I am hugely enthusiastic about it"
> "I think this is a fantastic opportunity for this department"

2- He shows limited ambitions relative to the protection of the Environment

- *Acknowledging a small budget:*

> "I know it is small, and I know it has a relatively small budget, but it has got, actually, in Chris Huhne, Greg Barker and Charles Hendry, it has got three very talented ministers"

- *Listing of worries*

He commonly quotes *climate change* in a list, not conferring much importance to each point:

> "And we'll have many subjects to discuss this morning: aid, development, climate change, the economic relations between our countries and also the importance of dealing with extremist terror in both our countries"

- *Imprecision of measures*

An Environment-related term with a high frequency in Cameron's speeches is the word *green*, conspicuously used in the example below:

> "We've got a real opportunity to drive the green economy – to have green jobs, green growth, and make sure that we have our share of the industries of the future"

In terms of relative frequency throughout the Cameron corpus, the word *green* (141 per million) actually ranks second after the term *energy* (182 per million) and before the term *climate* (108 per million). But the meaning of *green*, due to an extensive use in recent years, is ambiguous and therefore not very significant in political speech.

3- He constantly refers to global action

As opposed to Blair, Cameron does not isolate Britain's position as one major player regarding the protection of the Environment:

> "There's also a whole set of international issues where we meet together at the G8, the G20, the United Nations, climate change, where we should understand each other's positions and try and work with each other"

Indeed, studies have shown in the years 2000 that the major efforts of European nations in terms of CO^2 emissions for instance were neutralized or even undone, mostly by the development of the huge "BRIC" economies like China or India. The growth of the sectors of Transportation and Industry in those developing countries was a major offset to the efforts of European nations. In the same way as Blair, Cameron thus presents the UK as one actor among other major economic players:

> "Climate change does not respect borders: what is sown in one part of the world is reaped the world over. That's why we need global action, with all the major economies playing their part"

This global action does not only apply to developing countries but Cameron insists on the necessary collaboration of all European countries:

> "It is a working relationship. We face huge challenges today – from the conflict in Afghanistan to the economic crisis to climate change…and I am committed to working with President Sarkozy, working with France to tackle them"

Besides the support of developing and developed countries, Cameron often mentions the basic but fundamental commitment of British citizens.

> "Government departments have got to reduce their carbon emissions by 10%. That has to be done. You're going to drive it, you're going to lead it, but my ministers are going to help make it happen. And they're going to be helped in that by another involved party in terms of climate change, and that is the great British public"

This message was perceptibly linked to the framework of his "Big Society", as an attempt to empower people and reframe the role of government[7].

After investigating discourse argumentation and macrostructure, we will now focus on its microstructure that is the sub-sentence level and the lexis used in these political speeches.

---

[7] It should be remembered that the corpus for this study only covers the first year and a half of Cameron's mandate. All the above remarks therefore only apply to the period ending in December 2011.

Discourse and lexis

The lexis is understood in this study as the terms constructed from notions and organized schematically. We will not take into account the local enunciative situations, thus the lexis is not at the level of utterances[8]. Also, the lexical semantics used here does not encompass parallel syntactic structures as can be found in the literature[9].

In other words, we will look at the meaning of words and collocates to identify changes and relate them to the three prime ministerial corpora.

Presence and absence of terms

One of the most salient features of our corpora, even though it is only detectable through a meticulous manual examination of the lexis used by each PM, is the presence of some terms in the Blair corpus and their disappearance thereafter. The following set exemplifies the tendency to simplify or do away with some notions altogether:

	Blair	Brown	Cameron
air	49	0	0
atmosphere	14	0	0
chemical(s)	3	0	0
co2	36	0	0
greenhouse	74	0	0
litter	19	0	0
pollution	60	0	0
recycling	27	0	0

Fig. 4

Tony Blair uses terms related to one of the main themes of the 1970s up to 1990s by scientists and activists alike, namely air and soil pollution, as well as recycling.

[8] See Culioli in References

[9] See, for example, Levin, Beth and Hovav, Malka Rappaport in References.

These are not outright technical (except CO^2) and are meant to be easily understood and to resound in the everyday life of citizens, either on the ethical or economic level. The following examples from the Blair corpus illustrate this:

> "We need to tackle those issues which affect us all wherever we live. Tackling pollution, improving the environment and ensuring that our children will inherit a world at least as rich and diverse as our own"
>
> "Tougher powers for councils and neighbourhoods to tackle abandoned cars, litter [...], problems which plague the lives of many decent law-abiding citizens"
>
> "We need to [...] become a low-carbon, recycling economy"

Tony Blair seems to favor a bottom-up or "grassroots" approach where well-informed dwellers make concrete steps to help preserve their environment. On the other hand, Gordon Brown is confident in a top-down approach in which supranational bodies and national institutions resolve the more sweeping issues like climate change, whereby coverage is detrimental to precision; hence, there is no point in stating the details of a policy. David Cameron also urges citizens to act but, unlike Tony Blair, it is part of a transfer of governmental action to individuals or associations (see remarks on *Big Society* above).

Quantitative variation

Most of the time, only the relative frequency varies among the three premiers while meaning stays the same, suggesting expert agreement and lexical stability. This may however be deceptive at times, as we shall see.

The following set of terms exemplifies that a great variation mostly reveals individual thematic profiles.

	Blair	Brown	Cameron
climate	479	692	103
emission(s)	304	185	24
nuclear	51	322	55

Fig. 5

As noted earlier, Gordon Brown's stance on the subject of climate is rooted in the belief that climate should be equated with climate change (this very combination is used in about 90% of the occurrences) which is of course a simplification of the underlying scientific issues: in fact, hurricanes, floods, droughts etc. are probably pre-existing problems (not to mention the different global warming scenarios upon which scientists don't seem to agree).

Besides this, there is an obvious link with the notions of crisis (collocates such as *catastrophe/ic, chaos, drought* etc.):

> "I want to talk about how together we can forge and then legislate, for the first time, a truly global climate change agreement to save our planet from catastrophic climate change"

The overrepresentation of the term *nuclear* also calls for detailed explanation. Indeed, his mandate coincided with strong tensions between Western nuclear powers and Iran on their enrichment program which was seen as an attempt at producing nuclear weapons. This must be one of the reasons for the relatively frequent use of this term, but there is more to it: civil applications of nuclear power–as opposed to military ones– are promoted, suggesting that they may help in the global environmental problem of energy supply. So there's clearly a pro-nuclear standpoint on part of Gordon Brown:

> "We will seek an innovative partnership between industry, academia and government for further research and development to tackle the technical challenges involved in developing a proliferation-proof nuclear fuel cycle"
>
> "We will not meet the challenges of climate change without the far wider use of civil nuclear power"

Finally, as we have seen above, David Cameron is also interested in setting up international cooperation to curb climate change through emissions reduction while safeguarding our lifestyles by securing access to energy sources and developing technologies:

> "Clearly there's the climate change agenda, where we've got to get back on track, both nationally and internationally. And third, there is the issue of energy security, which I think is vitally important, which we need to do a huge amount of work on"

On the whole, he shares the views of his predecessors, even though we can note that he sometimes shows defiance against the validity of the current notions tied to climate issues:

> "So even if you're not 100 per cent certain about climate change, it's worth doing this for the insurance, for the risk avoidance"

Opposite uses

Departing from the largely monolithic terms of the Environment outlined above, we were able to zero in on a few terms that display antithetical uses.

Two such terms are *bus* and *railway*. Tony Blair is the only PM to use them in an environmental sense when his successors stick to the core features of the notion and insert them into a totally different context with either a positive or connotation:

Tony Blair's intention is to promote the clean buses of the future:

> "Environmentally-aware oil companies [...] are working with car manufacturers on new technologies, from developing fuel cell cars, to putting hydrogen-powered buses on the road in Europe"

He also castigates public railways as an alternative to road transport and calls for more private investment:

> "The railway system simply lacks the capacity to sustain our policies to expand both passenger travel and freight transport by rail, and to relieve road and air congestion. To deliver those policies, we must invest–with the lion's share of that investment coming from the private sector, in the public-private partnerships trailblazed by John Prescott"

Gordon Brown's intention is to fight terror in public transportation, which makes it more complicated because of security checks:

> "We too have learned the grief when lives are lost through terror on a bus or at the airport"

David Cameron questions welfare but calls for more public investment:

> "We're making cuts to the welfare budget which is hugely difficult so that we can fund big transport projects on our roads and railways"

Drifts in meaning

From a linguistic perspective, drifts in meaning are interesting because they demonstrate the speaker's will to shape representations of the world; in the particular case of political discourse, there's even a potential to change the world (be it symbolically or for real) through political action, even in a domain such as the Environment where human action seems to have little curative effect.

The set of terms below exemplifies the differences between the three prime ministers.

Clean: T. Blair uses this term for environmental purposes in 95% of the cases, for example 'clean energy', 'clean water' etc., as opposed to 40% for G. Brown and D. Cameron who prefer economic or social uses such as W*e must clean up the banking system* (G. Brown) or '*a clean press*' (D. Cameron). Of course, many statements involving *clean* are bold because there is often a downside that politicians choose to downplay; in the case of 'clean energy', it can be the cost of producing, maintaining or handling waste, for 'clean press' it is even more telling with hindsight (e.g. the *News of the World* scandal). But these caveats do not matter as long as a strong belief is forged.

Earth: T. Blair's use of it is synonymous with "Mother Earth" in about 70% of the cases and thus in environmental contexts whereas G. Brown's figure is about 25% environmental and even nil for D. Cameron; they both use it either as a synonym of 'worldwide' or in the expression 'what on earth'.

Waste: T. Blair's usage is 60% environmental e.g. *Most of our buildings waste more energy than they use because their heating, cooling and ventilation systems are inefficient.* On the other hand, G. Brown few uses are all related to nuclear waste, confirming his "atomic" leaning. D. Cameron's use of waste is overwhelmingly economic such as in *It is about opening up the system, being competitive and cutting out waste and bureaucracy.*

As previously noted, the drift from environmental to economic meaning observed mainly in G. Brown and D. Cameron's speeches is a major finding of the study and it rests with their author's vision of the world.

Conclusion

A logical termination of this analysis is to question the major differences between speeches. Do they seem to be time-related and interwoven with the period of the PM mandate? Do they convey political divergences between the Labour and Conservative Party, or are they simply linguistic or diachronic variations?

As preliminary information, it is important to mention that Environmental issues, whether they've been tackled by the Labour or Conservative Party, have never been a priority in British PM discourse. It has always been disregarded in favour of other national and international political matters.

Brown's mandate started with the world economic crisis. His discourse is therefore impacted by the international economic situation, let alone his personal leaning for those subjects.

Cameron's mandate is also tarnished by the crisis even though he starts putting new subjects on the agenda. However, the Economy has always been an important part of the Conservative ideology, the focus on economic matters is therefore not remarkable in Cameron's speech.

What is worth noticing though, is that Brown and Cameron's discourse have numerous common points. Indeed, as we have shown above, some terms like *clean,* CO^2*, litter, recycling, pollution and waste* have either disappeared from their speeches, or been used with different meanings after the Blair period. Some of these terms, ordinarily (and historically) reserved to the field of the Environment have been used with a more economic meaning.

As demonstrated in the analysis, it stands out that Brown's discourse is repeatedly closer to Cameron's than to Blair's. For some aspects, the differences are therefore not just political or linked to parties, but also correlated to time-periods, world events and PM personalities.

From a linguistic and diachronic point of view, we also worked on the vast *Google books*[10]database as a comparison, keeping in mind that it is not balanced. We analyzed the evolution of the following lexis on a British corpus since 1980: *litter, pollution, recycle, waste, emission, air, atmosphere* and *chemical.* A surge can be observed in the years 2000-2003 and a continuous downward curve thereafter. This actually reflects the trend in Brown and Cameron's discourse diachronically. Conversely, the use of CO^2, *greenhouse, climate* and *renewable* has been increasing over the last 15 years, not substantiating the tendency observed in PM speeches.

---

[10] 20 million books scanned in 2012. http://books.google.com/

These contradictory data show the linguistic complexity in determining inflexible general lexical tendencies. Besides this, the observation period is insufficiently long to conclude on the lexical emergence or disappearance of certain terms. The following perspectives will confirm and specify the necessary follow-up to this study.

Perspectives

The content of a PM discourse is a very enlightening source of information on a period, a personality, or a political ideology and legacy.

However, discourse analysis can only be comprehensive if speeches are actually compared. Indeed, one type of discourse will point out what is non-existent in another discourse, or vice versa.

One of the reasons for this is the will to leave a mark on History by means of differentiation: as we have shown above, Blair put environmental issues in the limelight. Thereafter, they have not been crucial in PM discourse. In order to draw more fine-grained conclusions about this linguistic evolution of terms, this study would need to be extended and completed by the analysis of John Major and Margaret Thatcher's mandates to try to separate diachronic linguistic trends from political doctrine. Putatively, a shift of paradigm within the Labour party on the Environment as opposed to a persistence thereof within the Conservative party could be a line of research, as well as the stance of the Liberal Democrats which was not explicitly included by using David Cameron's speeches.

Also, discourse analysis alone cannot account for the reception by the general public of facts about the state of the Environment and the corresponding actions taken by political powers. A sociolinguistic approach would be suited to complement this research, or even political science: parties nowadays make heavy use of polls which allows them to sense the subjects that matter to the general public. However, if there is no other loud and potentially discordant voice (e.g. experts, NGOs etc.), there is a risk that Environmental issues will never make it to the political level, especially if polls serve as a policy-shaping tool.

Bibliography

Biros, Camille, “Contribution à l’étude du discours environnemental : les organisations et leurs discours au Royaume-Uni.” PhD dissertation, Université Victor Segalen Bordeaux 2, 2011.

Culioli, Antoine, *Pour une linguistique de l’énonciation. Opérations et représentations, Tome 1*. Paris : Ophrys, 1990, p. 79.

Fill, Alwin and Mühlhäusler, Peter (ed.), *The Ecolinguistics Reader: Language, Ecology, and Environment*, London/New York: Continuum, 2001.

Fløttum, Kjersti, “A linguistic and discursive view on climate change discourse.” *ASp 58* (2010). Available online on 30 Nov. 2013. Paper version consulted on 09 Nov. 2012. DOI : 10.4000/asp.1793

Jalenques-Vigouroux Béatrice, “Dire l'environnement : le métarécit environnemental en question.” PhD diss., Université Paris 4, 2006

Levin, Beth and Hovav, Malka Rappaport, “Lexical Semantics and Syntactic Structures” In *The Handbook of Contemporary Semantic Theory,* ed. Shalom Lappin (Oxford: Blackwell, 1996), pp. 487-507.

Mayaffre, Damon, *Mesure et démesure du discours. Nicolas Sarkozy (2007-2012)*, Paris : Presses de Sciences Po, 2012.

Myerson, Georges and Rydin, Yvonne, *The Language of Environment: A New Rhetoric*, London : Routledge, 1996.

Sources of corpora

Number 10 Downing Street. The official site of the British Prime Minister's Office - http://www.number10.gov.uk/

Brown, Gordon, *The change we choose. Speeches 2007–2009*, Edinburgh: Mainstream Publishing Company Ltd, 2010.

THE LIBERAL DEMOCRATS AND THE GREEN CAUSE: FROM YELLOW TO GREEN

MURIEL CASSEL-PICCOT
JEAN MOULIN-LYON 3 UNIVERSITY

As heir to the Liberals and later the Alliance, the British Liberal Democrat Party, formed in 1988, has adopted "Libby", the flying bird of liberty, as their logo, and yellow as their "brand" colour. This choice is totally consistent: first the party's philosophy is centred on the freedom of the individual; secondly the colour yellow, commonly associated with the political ideology of liberalism in Europe, is a symbol of optimism,[1] which is congruous with their positive view of rational human beings and social progress.[2] Alongside these fundamental values, the basic principles of the Liberal Democrats are commitments to control and accountability of power, fairness, equality, pluralism, social justice, non discrimination and internationalism.[3] Inherited from the past, these tenets, among other factors, have enabled the party to emerge in British politics as a third force that has to be reckoned with. However, despite a policy of equidistance and a willingness to be "Neither Left nor Right",[4] the party, which used to be described as ideologically closer to Labour,[5] has now been in a coalition with the Conservatives since 2010. The position of the Liberal Democrats appears to be all the more complex as they have

---

[1] When the eye perceives the colour yellow, the brain releases more serotonin.
[2] Michael Freeden, *Ideologies and Political Theories* (New York: OUP, 1996), 198.
[3] The Liberal Democrats, *The Constitution of the Liberal Democrats.* http://www.libdems.org.uk (15 September 2012).
[4] Andrew Russell and Edward Fieldhouse, *Neither Left nor Right, The Liberal Democrats and the Electorate* (Manchester: MUP, 2005).
[5] The two parties worked in close cooperation before the 1997 general election, formed coalitions in Wales from 2000 to 2003 and in Scotland from 1999 to 2007.

acquired a strong reputation for being the greenest of the three main British parties.

Green thinking, the origin of which can be traced back to pantheism, pastoralism, and romanticism in particular, started to take shape in the 1960s and 1970s. Its aim is to create an ecologically sustainable society. Depending on its different branches, it dismisses anthropocentrism in favour of biocentrism and eco-centrism in various degrees.[6] Its core ingredients are limits to growth, a holistic approach to environmental issues, the respect of diversity and biodiversity, ethical standards including the defence of animal rights and a sense of overall responsibility towards future generations, social justice, and a preference for decentralisation and local organisation.[7] The Green ideology shares a number of common threads with conservatism and socialism, such as an attachment to the land and a rural way of life, a defence of the natural heritage for the former, a condemnation of capitalism, a search for radical social change, and a long-term interest in humanity. In practice however, it has been more commonly associated with the left rather than the right.

In this context, the Liberal Democrats do not only need to differentiate themselves from the Conservative and Labour Parties, while finding common ground in case of political necessity, but they also have to strike a balance between their "liberal adherence to [...] non-interference in people's choices and lifestyles and their desire to limit the environmental consequences of those choices."[8] Conrad Russell has notably discussed the argument when, after defining green liberalism as "the outmost layer of the onion of Liberal philosophy," he defended the Liberal Democrat commitment to sustainable economic growth. Thus the combination of liberal and green thinking, which is not described as a potential contradiction in terms, leads to the view that individuals pursuing their own best interests can create a sustainable society.

If Green Liberalism is thus justified, how is it pursued by the Liberal Democrat Party? This paper aims to chart the green trajectory of the champions of liberty and equality, to assess how green they have become over the years, and to analyze to what extent the greening process has been a source of cohesion and/or division. Has the party managed to reconcile the harmonious colours of yellow and green, which sit next to

---

6 Duncan Brack and Ed Randall, *The Dictionary of Liberal Thought* (London: Politico's, 2007), 106.

7 Andrew Heywood, *Political Ideologies* (Basingstoke: Palgrave, 2007), 255-280.

8 Conrad Russell, *An Intelligent Person's Guide to Liberalism* (London: Duckworth, 1999), 99.

each other on the colour wheel but might not be so close in the political sphere?

Taking up the Green Cause

Humanity's concern for its impact on the environment is not new. In Roman times there were complaints that deforestation caused soil degradation.[9] In the reign of James I, there were worries about dead animals being thrown into London's water supply. These problems did not give rise to public activism. It was only in the wake of the Industrial Revolution, when health and pollution issues were raised that contemporary environmentalism emerged. And it was only in the 1960s and 1970s that people started to realise that the degradation of the environment and the depletion of resources were the inevitable consequences of Western modern social and economic patterns.

The Challenge to Man's Future, written by Harrison Brown in 1956, was the first book about the uncertainties facing the world. At the time, the British economy was thriving, and the Liberal Party, like its rivals, considered production, trade, and consumption as positive elements that had to be encouraged.[10] During the 1960s, Rachel Carson's *Silent Spring* and Lewis Herber's *Our Synthetic Environment* exposed the dangers of pesticides and denounced the role of the chemical industry, while other works dealt with overpopulation and starvation. The early 1970s witnessed the release of three major publications: the Club of Rome's *Limits to Growth,* which explored the possibility of global sustainability;[11] *The Ecologist's Blueprint for Survival*, which recommended a radical restructuring of modern society based on decentralisation; and E.F. Schumacher's *Small is Beautiful*, a criticism of Western economics and an advocacy of the empowerment of people. These publications had a real impact on some sections of the population, including the Liberal Party, and their perception of the environment. Indeed in 1972 the Liberals produced a report about population growth, pollution, economic growth and the finite resources of the world. The study led by Inga-Stina Robson can be considered as the real beginning of the greening of the party.

---

[9] Duncan Brack and Ed Randall, *The Dictionary of Liberal Thought*, 105.

[10] Tony Beamish, "The Greening of the Liberals?" *Journal of Liberal Democrat History* 21 (1998-1999): 15-19.

[11] The study was based on the computer modelling of five variables: population, industrialisation, pollution, food production and resource depletion. It presented several scenarios for the future of the world's global system, including collapse and stabilisation.

Meanwhile, the first national Green Party in Europe, PEOPLE, was founded in Britain in 1973.

This first proof of green awareness among the Liberals was not echoed by any concrete measures in their manifestos. In 1970, the only reference to the protection of the environment was: "We are only just waking up to the dangers of pollution and the damage we have done to the environment and the quality of life". The 1974 (February) programme showed some slight progress, acknowledging the enormous damage done to society by the pursuit of unlimited growth and the obligation for the party to provide a safe environment. A dozen lines insisted on controls of the use of all potential pollutants, recycling, rail transport, the preservation of natural resources, and action at international levels. The 1979 manifesto reflected a growing "concern to bring to bear an environmental perspective across the whole range of government policies." A whole paragraph expressed the party's beliefs in a wiser use of resources, war on pollution, the responsibility towards future generations, and a sound basis for economic and social activity.[12] A series of proposals was made, some of which sound familiar today: increase research and development in sustainable sources of energy, discourage the use of private cars in cities, and make polluters pay the price of their pollution. In like manner, the 1979 "Liberal Assembly declared that economic growth as conventionally defined was neither indefinitely possible nor desirable."[13] That period marked a second step towards a greening of the party.

While retaining the same approach in 1983, the Alliance's manifesto relegated the issue to the sidelines. It appeared in one sub-part entitled "The Environment" in the third part *Creating One Community*. The document did not support any new or more precise developments. However, for the first time it clearly stipulated the strong link between the economy and the natural world: "There can be no healthy economy without a healthy environment." The statement was well illustrated in the second Alliance's manifesto with one part focusing on "Green Growth." It proved the party's faith in expansion and its will to set "the right balance between present consumption and future investment and [to ensure] that economic development is sustainable and environmentally responsible." The length of the section evidenced renewed interest for the

---

[12] The paragraph "Conservation of Resources and Environment" was divided into four areas: Energy and North Sea Oil, Transport, Food and Agriculture, and Safeguarding the Environment.

[13] Duncan Brack, "Liberal Democrat Policy" in *The Liberal Democrats*, ed. Don MacIver (Hemel Hempstead: Prentice Hall, 1996), 107.

green cause while the Chernobyl disaster on 26 April 1986 provided a further incentive. For the first time, a real set of precise measures was presented including the setting up of a new Department of Environmental Protection, a reiteration of the "polluter pays" principle, higher standards of insulation in homes, and reforestation programmes. Meanwhile the Liberals also continued to advocate the development of a modernised coal industry, research into nuclear fission power and the disposal of nuclear waste. Hence in 1987 the greening of the manifesto was relative.

The 1992 general election saw the arrival of the Liberal Democrats, who produced a programme that "had a good deal of the spirit of classical liberalism" but also placed great emphasis on environmental policies.[14] This pivotal moment came to the newly born party as the result of a shock. Indeed the descendants of the Liberals had suffered a resounding defeat at the 1989 European elections when they came fourth behind the Green Party in every single seat in Britain. Neither party gained seats, but the latter collected 15% of the vote share while the former received only 6%. Paddy Ashdown, the party leader, was very worried about the Green phenomenon.[15] The event was "a particular blow in view of the effort which the Liberals had made since the 1970s to focus upon environmental issues."[16]

> Our defeat was especially difficult for us to swallow, because we hadn't so much missed the boat as found that the boat we had built was being skippered by somebody else [...] Our belief was that people who understood environmental issues and care about them, would vote for us, because we felt the same. That led to a degree of complacency.[17]

As a consequence the party (like its rivals) "took energetic steps to establish their environmental credentials" even if in those years the issue was still low in salience compared to health and taxation.[18] The Liberal Democrats could transform their awareness of the environment into political awareness. To what extent did it show in their subsequent manifestos?

---

[14] Roy Douglas, "Classical Liberalism" in *The Political Thought of the Liberals and Liberal Democrats since 1945*, ed. Kevin Hickson (Manchester: MUP, 2009), 25.
[15] Paddy Ashdown, *The Ashdown Diaries 1988-1997* (London: Penguin, 2000), 50.
[16] David Dutton, *A History of the Liberal Party in the Twentieth Century* (Basingstoke: Palgrave, 2004), 270.
[17] Charles Kennedy, *The Future of Politics* (London: HarperCollins, 2000), 52-53.
[18] David Butler and Dennis Kavanagh, *The British General Election of 1992* (London: Macmillan, 1992), 81.

Ever Greener

The Liberal Democrat Party considered that they were well placed to address green issues because the quality of the environment was actually their own territory in the first place. However it did not come so naturally to them, as evidenced by the above context, not to mention the 1992 Rio Summit. It offered them a new area of debate, which as policy innovators they should have been very efficient at embarking on. Secondly, even if the Liberal Democrats' manifestos declared their green intentions, their leaders' writings were rather timid on the subject. Though Paddy Ashdown wrote a book entitled *The Environment* in 1990, and declared at the end of the 1992 campaign that "these are the issues which this election should have been about. […] About the threat of pollution, and the measures we can take to deal with it now,"[19] he devoted little space to the problem in his *Diaries*. Charles Kennedy himself confirmed that when he became involved in politics "there was little room in [his] conscience for matters such as pollution and global warming"[20] and that the issue was partly forced on him and the party by the electoral disaster of 1989. Furthermore Chris Bowers commented:

> the political noses of Campbell and Charles Kennedy had told them that the environment was important, but I never got the impression it was something they really felt. […] Clegg certainly recognises the environment's importance […] But in the run-up to the general election he left a lot of his senior party wondering whether he really got it.[21]

Liberal democrat leaders seem to have intellectually understood the green cause and they know that they cannot ignore it. But they have had difficulty dealing with it. To some extent, the environment has imposed itself on them not only as the latest political challenge to be seized but also as an electoral strategy to remain in third place on the British political stage. Furthermore the leaders have also had to respond to internal pressure from the members and from the Green Liberal Democrats. The latter hold strong views on environmental issues according to a survey carried out in 1998-1999.[22] The results showed that

---

[19] David Walter, *The Strange Rebirth of Liberal England* (London: Politico's, 2003), 162.

[20] Charles Kennedy, *The Future of Politics*, 51.

[21] Chris Bowers, *Nick Clegg, The Biography* (London: BiteBack, 2011), 4 & 203.

[22] Paul Whiteley, Patrick Seyd, and Antony Billinghurst, *Third Force Politics* (Oxford: OUP, 2006), 30-31.

81% of the members were opposed to the building of houses in the green belt, 79% believed that modern farming methods harmed the countryside, 55% disapproved of the development of nuclear central stations, 53% thought that car users should pay higher taxes, and 86% were in favour of paying more for a radical environmental policy. The former as "an associated organisation" and major campaign group are important players who aim to work continually to make their party the 'greenest' party in the UK, to strive to be its environmental conscience, and to develop links with the wider green movement. Hence the successive leaderships of the party have had to take into account internal and external forces while being aware of the difficulty in selling green policies to voters. In this context, how has the greening of the party shown in the manifestos?

Visually speaking the party has adopted the colour green in its original manifestos since 2001. Under Paddy Ashdown's leadership, the 1992 and 1997 manifestos were a warm yellow; both electoral programmes devoted a chapter to environmental issues that respectively ranked second and third in the lists of sections.[23] Things changed under Charles Kennedy's leadership. In the 2001 manifesto, the message from the leader did not set the protection of the environment as a priority because it had become a backdrop to all liberal democrat policies.

> All our policies have a green dimension. So there is an environmental section in every chapter, a green thread binding together all our thinking. Without steps to preserve our planet for future generations, none of our other policies would have much purpose.

Using recurrent green boxes, the party could boast a widening of its green policies in a soft style. The 2005 manifesto adopted the same green thread but combined it with a reiteration of the party's green pledges in Chapter 11. In 2010, the tradition of an environmental conducting theme was maintained but in the form of more discrete green tags. For the first time the front cover was not solely yellow. Against a white background four yellow, light green, light blue and aqua strips bore the party's priorities, among which the environment coupled with employment ranked third.[24] The idea of a chapter entirely devoted to green issues was abandoned and references to them only appeared in the subtitles of three

[23] *Britain's Environment: Environmental Protection and Conservation*, 1992. *Environment "Which Party is Serious about Making our Environment Cleaner and Safer?"*, 1997.

[24] "A fair future, creating jobs by making Britain greener."

sections.[25] Thus at first sight the manifestos have turned greener but their greening has included mitigating factors. This interpretation is confirmed by the quantitative analysis of party programmes carried out by the Manifesto Research Group.[26] Counting the number of sentences referring to policy positions, the findings are as follows:

Table 2.1 Party Positions

G.E. Year	Environment vs. Other Party Positions		Closest Policy Positions	Closest Policy Position	
	Share %	Rank		Rank	Share %
1992	11.5	1	Welfare State Expansion	2 3	7.9
1997	8	2	Welfare State Expansion	1	9.5
2001	7.4	2	Political Authority Welfare State Expansion	1 3	11.4 7.3
2005	9.3	3	Political Authority Non Economic Demographic Groups	1 2	15.8 10.7
2010	7.4	2	Political Authority Non Economic Demographic Groups	1 3	10.2 6.3

Data compiled by the author.

Green issues are quantitatively at the top of the Liberal Democrat agenda but they have tended to recede in terms of share and rank. At this stage, a qualitative analysis of the contents will help understand whether the widening has been accompanied by a deepening of the party's green arguments.

---

[25] "Your Job", "Your World", "Your Community."

[26] Manifesto Project Database, https://manifestoproject.wzb.eu (15 September 2010).

A Deeper Shade of Green

Assessing the change in the green shades of liberal democrat policies as expressed in their original manifestos[27] means evaluating whether the party has become more aware of environmental considerations. This implies an examination of the party's suggestions for generating economic activity while preserving environmental quality and using natural resources efficiently. Combining these three aims covers reducing waste and pollution, managing energy use, preserving ecosystems, improving human well-being, and promoting social inclusiveness.[28] For political players, achieving these goals involves creating financial and economic incentives, fostering innovation, encouraging investment, establishing standards, and imposing regulations for production, distribution, and consumption. The following assessment of these elements is not exhaustive. However it is representative of the major issues raised through a variety of recommended solutions.[29] Comparing specific methods is a tangible way of highlighting the intensity of the party's greening.

Firstly, as climate change is considered a priority, one of the most telling tools is the setting of emission targets. In their five manifestos the Liberal Democrats have maintained their intention of reducing CO2 emissions using the base year level of 1990.

Table 2.2 shows that, when compared to what has been happening since 1990, the party has definitely ambitious and relatively stable targets. Another solution to diminish global warming is to establish targets for the production of clean energy. In 2001 the Liberal Democrats declared that they wanted to have a minimum of 10% of the UK's energy to be generated from UK-based renewable energy sources by 2010. The figure was overly optimistic as in 2010 renewables (excluding nuclear power) accounted for only 0.75% of the domestic production of primary energy.[30] In 2005 and 2010 the targets were downgraded, as they no longer dealt with energy but electricity. In 2010, the percentage of electricity from renewable sources in the United Kingdom was 7.4% (8.7% in 2011) while the party's targets were 20% and 40% by 2020 respectively in 2005 and

[27] The study focuses on original manifestos and not on their supplements when they exist.

[28] UNEP, *Green Economy*, http://www.unep.org/greeneconomy (15 September 2012).

[29] See the Green Alliance's strategic areas of research.

[30] DECC, *Energy Trends 2012*, 9. http://decc.gov.uk (15 September 2012).

2010.[31] This example shows that the content and scope of the measures are subject to variations that blur the degree of greenness of the party. Similarly between 1992 and 2005, the Liberal Democrats were committed to phasing out nuclear-power stations first by 2020 and then when they would reach the end of their safe operating lives. In the 2010 programme, the issue of nuclear waste, which had appeared earlier and remained a topical subject, was neglected, even if the construction of a new generation of nuclear plants was still rejected.

Table 2.2 Liberal Democrats' CO2 Targets

Manifestos	Reduction Target %	Dates	Number of Years since Base Year (1990)	Theoretical Reduction Rate per Year %
1992	30	By 2005	15	2.00
1997	30	By 2012	22	1.36
2001	20	By 2010	20	1.00
2005	"Targets Needed"	NA	NA	NA
2010	40 100	By 2020 By 2050	30 60	1.33 1.66
UK CO2 Emissions	*16*	*1990-2010*	*20*	*0.80*

Data compiled by the author.

Setting quantitative objectives is an attempt to produce outcomes that are achieved through regulations. These legal, governmental decisions create and constrain rights and duties, impose controls and sanctions, and allocate responsibility. The first three Liberal Democrat manifestos clearly recommended strong governmental interference such as: forbidding air and soil pollutants, enforcing pollution controls, prohibiting experiments and testing on animals, penalizing polluters, protecting the countryside and greenfield sites, imposing building standards, improving the regulatory system of public transport, and phasing out nuclear power stations. However, despite the reiteration of promises to cancel plans for airport extensions in the South-East (Heathrow in particular),[32] there has been a marked decline in clear indications of government interference since 2005. This was exemplified by the absence of terms like "banning" and the

[31] Ibid., 55.

[32] The pledge was made again at the party's 2012 Brighton conference.

decrease in the numbers of projected laws and bodies.[33] Showing less need for recourse to legislation and rules does not particularly mean that the party is potentially less green but conversely demonstrating less interest in a tool that acts both swiftly and decisively is not a proof of greening.

Taxation, defined as a financial contribution imposed on an individual or legal entity by a government, is another tool at the disposal of politicians. Ecological tax against products and practices is intended to curb methods that harm the environment, to promote environmentally sustainable activity and behaviour via economic incentives so that those involved will feel responsible for their actions towards society. Such re-pricing is intended to reduce environmental impact. From 1992 to 2001, the Liberal Democrats argued for shifting the burden of taxation to pollution and depletion of resources, while specifying that their recommendation meant taxing differently, not more. In 2005 and 2010 the leitmotiv changed and the references to green tax reforms were less precise. Between 1992 and 2005, the party notably advocated the replacement of the Climate Change Levy by a Carbon Tax, lower or no taxes (Vehicle Excise Duty) on less polluting vehicles, a modification of the aviation tax system including a levy on freight (from per passenger duty to per plane duty), and the introduction of congestion charges in cities.[34] In 2010 the Liberal Democrats' green taxation programme was reduced and focused on aircrafts and flights essentially. The limitation could be partly accounted for by the progress made by the previous governments towards environmental taxation such as increasing the landfill tax.[35] However, it could also be argued that former proposals could be retained, the charges increased, their collection extended to other fields, and their use made more effective.[36] In their 2010 manifesto, the Liberal Democrats did not innovate in the field of green taxation at a time when the issue needed clarification and an implementation plan. If restricting the scope of environmental taxation might not necessarily imply that the party is less green, it nevertheless proves that it deprives itself of means and schemes that are accused of hampering competitiveness but that are also

---

[33] Departments of state, commissions, agencies, offices, and tribunals.

[34] Congestion charges are also considered as economic instruments.

[35] European Environment Agency, "Environmental Taxes: Recent Developments in Tools for Integration", *Environmental Issues Series* 18, 2000, 75.

[36] Environment Audit Committee, *Report: Budget 2011 and Environmental Taxes.* http://www.parliament.uk (15 September 2012).

well-known for curtailing harm to the environment[37] and for having a positive impact on innovation.[38]

Moving away from penalising taxation means moving closer to encouraging incentives, market-based solutions, and investment. Indeed since 1992, the Liberal Democrats have repeatedly defended expenditure in renewable energy research, public transport (high-speed trains, reopening of disused rail tracks, city trams), and energy conservation (insulation). The 2010 manifesto even made a marked difference as its proposals were detailed and quantified.[39] A total of £2.44m was meant to be spent on insulating private homes (a one-year Eco-Cash Scheme) and public buildings, bringing old houses into use, replacing old polluting buses, and refurbishing shipyards to manufacture offshore wind turbines and marine renewable energy equipment. Other non-quantified financial contributions such as the transformation of the electricity network, the development of environmentally friendly biogas, and the stimulation of innovation at all levels, remained part of the agenda. Supporting heavy environmental investment to turn it into a viable growth scheme illustrated the Liberal Democrats' darkening green position. These financial proposals were in line with the provision of incentives such as cutting rail fares for passengers and the adoption of market-based solutions like cap-and-trade schemes. The party advocated the use of tradable emissions licenses as early as 1992. After their implementation, they recommended sectorial (aviation and shipping) and international extensions of the system, as well as tougher ceilings. A third example in 2010 was the creation of an Infrastructure Bank, which would not be nationalised and which would tap into the demand from investors such as pension funds, to direct private finance for essential rail projects.

The study of some of Liberal Democrats' pre-eminent green promises and adjoining methods tends to show that the greening of the party over the years has been relative and that the intensification of the colour has to be qualified. Firstly, the recommendations advocated were quite green from the start. Secondly, the party has been eager to maintain its green thread over the years. Thirdly, the refinement of the proposals, partly explained by the implementation of green measures, political experience, and scientific advances, can be interpreted as a sign of increased greenness. Fourthly, it can be hard for the party to enhance the greenness

---

[37] European Environment Agency, 49.

[38] European Environment Agency, "Environmental Tax Reform in Europe", *EEA Technical Report*, 17/2011, 22-23.

[39] The 1997 liberal democrat manifesto came with a costing commitment and a supplement.

of its agenda, as environmental issues are global, pressing, and unavoidable for all political parties. Fifthly, given that green proposals can be difficult to sell to voters, the Liberal Democrats have chosen out of conviction, through strategy and/or pragmatism to make them more palatable by reconciling them with economic supply-and-demand solutions, which in turn can be seen as a proof of governance competencies. However such a choice has led the party to disfavour an integrated approach, disregarding certain tools while focusing on others. In other words, the Liberal Democrats have given greater priority to the adjustment of their green shade to the current global economy than to an unprecedented deepening of the hue. In view of the above, two questions are raised. From a general point of view, what is the degree of compatibility between the protection of the environment and the economic model of global competition mainly allowing room for market-based solutions? The real issue is not so much the traditional economic either/or debate of sustainability and development as the moral reconciliation between the freedom of rivalry according to market rules, the constraint imposed by the principles of renewability, and the sharing of a circumscribed planet with limited resources. From a more specific point of view, have the green choices of the Liberal Democrats been as unifying a thread as yellow liberalism?

Green Concord vs. Green Discord

The Liberal Democrat Party has been overtly green. Its reputation started at a time when campaigning for the environment was not yet fashionable.[40] More recently, the 2007 *Green Standard Report* concluded that:

> The Liberal Democrats have scored well against the majority of our tests for environmental leadership. The party has clearly grasped the challenge of climate change mitigation and has developed an ambitious set of policies to address it.[41]

Friends of the Earth named the party's 2010 manifesto the most ambitious green programme of the three main parties.[42] The information

---

[40] Ed Davey, "Liberal Democrats Call for Coalition to be Greenest Government Ever", http://www.libdems.org.uk (15 September 2012).

[41] *Green Alliance, How Green our Parties are ? The Green Standard Report, 2007,* 20.

[42] Friends of the Earth.

was based on the Green Agency's 2009 *Common Cause: The Green Standard Manifesto on Climate Change and the Natural Environment.* The positive assessment of the party's commitments by leading British environmental groups has largely contributed to enhancing its green reputation even if it also rests on solid ideological grounds.

From a conceptual point of view, the doctrine of sustainability fits into traditional Liberalism and particularly Locke's theory with its central theme of trust. Power, which rests on the consent of the trustors, is conferred upon the trustees whose duties are to "hand on the inheritance to the next generation in as good a shape as he can leave it."[43] It is also consonant with John Stuart Mill's concerns with limits to resources, increasing personal consumption, and quality of life issues.[44] More generally individual freedom, which is one of the core values of the party, can be largely threatened by pollution and environmental changes.[45] It then follows that, if the environment is subject to envy and competition, succeeding in protecting the natural world and delivering environmental goods is the necessary precondition for securing equality of opportunity, social justice, and fairness. Furthermore in the sense of conscious thought and planning that allow the accomplishment of legitimate goals promoting human welfare,[46] rationality, another fundamental liberal assumption, agrees with the importance attached to scientific research about the environment. Last but not least, the Liberal Democrats' internationalism and communalism are totally congruous with problems that cannot be solved within national frontiers only and that also require practical actions about practical life at local levels. The reconciliation of liberal and green concerns can thus be seen as a source of ideological cohesion within the party.

Another sign of the Liberal Democrats' unity around environmental values is the convergence of views found in the *Orange Book* and *Reinventing the State, Social Liberalism in the 21st Century*, respectively written by the economic and social strands in the party. Providing political orientations, both works devote pages to environmental problems and solutions and despite differences in tone and messages, the extent of the

---

http://www.foe.co.uk/resource/press_releases/manifesto_assessment_election2010_26042010.html (15 September 2012).

[43] Conrad Russell, *An Intelligent Person's Guide to Liberalism*, 104.

[44] Duncan Brack, Ed Randall, *The Dictionary of Liberal Thought*, 109.

[45] Richard Grayson, in Kevin Hickson, *The Political Thought of the Liberals and Liberal Democrats since 1945*, 59.

[46] Paul Thagard, "Rationality and science" in A Mele & P Rawlings eds., Handbook of rationality (Oxford: OUP, 2004), 378-379.

disagreements between the two can be seen as limited. On the one hand, it was clear that, to ensure sustainability, both books recommended the use of market mechanisms, which would harness economic forces. But on the other hand, Chris Huhne's demonstration in *Reinventing the State* was less blunt, more detailed, and more balanced than Susan Kramer's argumentation in *The Orange Book.* Well versed in banking and finance, she started by establishing the unmanageability of regulation and taxation in order to emphasize the pragmatism and flexibility of trading schemes.

> Ultimately the regulation route threatens to be unmanageable [...] the taxation system is often an inadequate tool because it cannot deliver guaranteed outcomes [...]. We prefer a market-driven solution.[47]

Conversely Chris Huhne first assessed the practicality of economic instruments so that he could then focus on regulation and institutional reform. More recently at a Social Liberal Forum on 18 June 2011, he declared that none of the party's aims concerning energy efficiency could "be achieved through the market by itself" and that smart regulation can definitely help competition and innovation."[48] At this stage, we may wonder: had Chris Huhne with his spinach green statement of aims won the leadership election, would he have addressed sustainable development challenges in the same way as under Nick Clegg's guidance and vice-premiership? Indeed when he resigned from his post as energy secretary,

> Environmentalists expressed dismay at the loss of Huhne from the heart of government. Greenpeace said he had been "a vocal advocate for the green agenda in a government whose green credentials are looking more than a little tarnished".[49]

In view of the above considerations, the number of green proposals to be found in the 2010 manifesto and their contents are reflections of the *Orange Book* orientation.

This conclusion is relatively at odds with the findings of a *Liberal Democrat Voice*'s survey of its members carried out in 2011, which reveal that

---

[47] Paul Marshall, David Laws, *The Orange Book, Reclaiming Liberalism* (London: Profile Books, 2004), 176, 178, 189.
[48] Liberal Democrat Voice, http://www.libdemvoice.org/chris-huhne-we-need-smart-regulation-in-the-energy-sector-24487.html (15 September 2012).
[49] Government on line, http://www.government-online.net/chris-huhnes-successor-faces-clash-as-tories-attack-wind-farms-spending/ (15 September 2012).

> Almost two-thirds of members (64%) identify themselves as 'Social liberal', compared with 35% happily self-identifying with 'Economic liberal': given these are the two terms with greatest currency at the moment to describe the different 'left/right' wings of the party.[50]

In addition, 41% of members described themselves as green, which means that while not a majority, they represented a significant part of the group. It logically follows that in order to avoid potential discord the party should show coherence between its green credentials and its practical recommendations. Real actions carried out at local levels as in Sutton should not be overshadowed by positions and political difficulties at national level.[51] Although the coalition programme has been described as remarkable for its range of green measures, it proves disappointing in a difficult economic context. The implementation of the Green Deal,[52] officially to be launched on 1 October 2012, has already been subjected to postponements for commercial buildings and repayment mechanisms.[53]

Similarly the subsidisation of wind-farms, which caused rows between Chancellor George Osborne and senior Liberal Democrats earlier in 2012, has been cut while tax relief has been introduced for gas companies.[54] In this respect Ming Campbell has declared

> Liberal Democrat voters, and in particular Lib Dem activists, will not be content if there is any rowing back on the green agenda. Commitment to the environment has an iconic place in the Lib Dem approach and if we were seen to water that down, publicly and privately, things could get very tough.[55]

The same argument already applied in 2011 when Jonathan Porritt, former head of the Sustainable Development Commission, accused the Liberal Democrats of timidity and the coalition of delivering little.[56] Other

---

[50] Liberal Democrat Voice. http://www.libdemvoice.org/how-lib-dem-members-describe-their-political-identity-liberal-progressive-and-social-liberal-top-the-bill-23928.html (15 September 2012).

[51] Danny Alexander, *Why Vote Liberal Democrat* (London : BiteBack, 2010), 88-90.

[52] DECC, *The Green Deal, A Summary of the Government's Proposals*, 10D/96, p. 21.

[53] http://www.endsreport.com/33581/green-deal-delayed-for-commercial-buildings (12 September 2012).

[54] The Independent, "Wind Farm Subsidies to Fall by 10%", 25 July 2012.

[55] Toby Helm, "Chris Huhne's Successor Faces Clash as Tories Attack Windfarms Spending", The Guardian, 4 February 2012.

[56] Jonathan Porritt, *The Greenest Government Ever: One Year One, A report to Friends of the Earth,* May 2011.

types of dissensions and contradictory information over green policies could prove harmful to the party when major personalities shift from opposition to support of new nuclear power stations[57] or when disagreement at the top of the party evidence a lack of cohesion over environmental issues.[58]

The defence of a green agenda, which is by definition rather radical, by a traditionally yellow organisation has not been particularly easy. However, green is a unifying value for the Liberal Democrats because it is one of their best competitive advantages, a real issue of differentiation for 2015. Simultaneously given the economic and international contexts, it is probably the most difficult "unique selling proposition" to implement.

This paper has not been an attempt at evaluating whether, from an insider's point of view, the Liberal Democrats have genuinely and truly grown greener. It has attempted to assess the greening of a yellow party from the various ways it has communicated its environmental stands, orientations, and policies. The matching of the two harmonious colours came at an early stage and from the start environmental recommendations have been quite strong. With time, the party has widened its environmental theme but deepening its shade has been more difficult and constrained.

The primary hue has kept its distinctiveness notably through a focus on market-based solutions. However the secondary colour will be a differentiating trademark in the future, provided the party's green stand is firm. As the pace and effectiveness of the measures taken nationally and internationally appear to be slow and limited by scientific standards, it follows that the party may well be at an advantage. Conversely the complexity of environmental issues is such that it requires considerable political courage: a will to resolve paradoxes, a considerable effort to promote real global governance, and especially the introduction of changes in the dominant social model.

---

[57] Chris Huhne, "Why the Future of Nuclear Power will be Different", Speech to the Royal Society, 13 October 2011, http://www.decc.gov.uk (15 September 2012).

[58] The Guardian, "Why is Vince Cable Fighting Green Liberal Democrat Policies ?", http://www.guardian.co.uk/environment/damian-carrington-blog/2011/sep/19 (15 September 2012).

Bibliography

Brack, Ducan, Grayson, Richard & Howarth, David. *Reinventing the State, Social Liberalism for the 21st Century*. London: Polico's, 2007.

Carter, Neil, 'Party Politicization of the Environment in Britain', *Party Politics* 2006 Vol 12: 747-767.

Liberal Democrats, *Changing Britain for Good. The Liberal Democrat Manifesto 1992*. Dorchester: Liberal Democrat Publications, 1992.

—. *Make the Difference. The Liberal Democrat Manifesto 1997*. London: Liberal Democrat Publications, 1997.

—. *2001 Liberal Democrat General Election Manifesto. Freedom, Justice, Honesty*. London: Liberal Democrat Publications, 2001.

—. *Liberal Democrats. The Real Alternative*. London: Liberal Democrat Publications, 2005.

—. *Change that Works for You: Liberal Democrat General Election Manifesto 2010; Building a Fairer Britain,* London: Liberal Democrat Publications, 2010.

Marshall, Paul & Laws David, *The Orange Book, Reclaiming Liberalism.* London: Profile Books, 2004.

Russel, Conrad, *An Intelligent Person's Guide to Liberalism.* London: Duckworth, 1999.

PART IV:

THE ENVIRONMENTAL ISSUE IN SCOTLAND AND WALES

GREEN SPEECH IN WALES: A POLITICAL ISSUE

STÉPHANIE BORY
JEAN MOULIN-LYON 3 UNIVERSITY

In an article entitled "Public Sector Leadership on Sustainability" and published in July 2010, Peter Madden, the director of Forum for the Future, a British nongovernmental organisation collaborating with business and governments, explained that most public sector organisations in the United Kingdom willingly accepted the environmental imperative with policy, strategy, resources, delivery all lined up to achieve sustainable development, but that among them very few had developed a truly joined-up approach. He added that the Welsh Government, proud of its green agenda, had asked him two years before: "What would it look like if we were to become the leading public sector organisation on sustainable development?". Indeed, the National Assembly for Wales (NAW) is one of the three administrations in the world endowed with a legal obligation to promote sustainable development, and became in October 2010 the most sustainable public sector organisation in the United Kingdom during the Sustainable Facilities Management Awards 2010. Besides, a survey commissioned in 2011 by the Welsh government to find out Welsh views and attitudes to sustainability and climate change revealed that 72% of people agreed that these issues should play a key role in decision-making in Wales. After twelve years' existence, the Welsh institutions have recourse to environmental issues to assert and acknowledge Wales' specificity, or even her superiority, at least in Europe. Even if the Wales Green Party is quite weak in the country, as illustrated by the following remark made by Gareth Clubb, chairing Friends of the Earth (FoE) Cymru, during an interview he gave to the author in March 2012: "Wales Green Party doesn't exist – it's the England and Wales Green Party. That is a problem for the Green Party's credibility". How has a small peripheral government achieved such an outcome? How can recent institutions present themselves as a model within the United Kingdom?

This article will first aim to remind that environmental issues have represented a nationalistic niche in Wales for several decades since questions of identity and the protection of the environment have been deeply interconnected. At the end of the 20th century, the Welsh voice, thanks to the green speech, seemed at last to be heard, both on the British and European stages. And yet, turning words into concrete actions rapidly turned out to be difficult for the new institutions. Welsh people then had the feeling that the Assembly, in charge with defending their interests, including the landscape, did not respect its commitments and was eventually unable to stick to its word. There was thus sometimes no possible communication between people and power. As a result, Welsh people, caring for their environment, have turned to other organisations so that the green speech, representing a real political issue, is more and more delivered by environmental nongovernmental organisations (ENGOs), lobbies which make themselves heard, even more than politicians themselves. Are they to become the main voice dealing with the green speech in Wales?

Environmental stakes: a nationalistic niche

First of all, some events have marked the Welsh psyche, such as the building at the end of the 1950s of a nuclear power station in Trawsfynydd, right in the middle of the Snowdonia national park set up in 1951. The scheme divided people, torn between their will to protect the landscape and the economic necessity. Even if they were worried by the plan, they supported a job-creating sector while many environmental organisations denounced the scarring of the landscape. Welsh people realized then that such a decision was made in London, not in Cardiff, even more so since the scheme in Trawsfynydd coincided with that of a reservoir in Tryweryn, a village located just a few kilometres away from the site found out for the power station. It implied drowning a whole valley and so moving a 70-people Welsh-speaking community to provide factories in Liverpool with water. It caused deep resentment and gave way to a fierce opposition in Wales, especially from Plaid Cymru, the nationalist party, but it nevertheless materialized, being supported both by the Liverpool Corporation and the British Parliament. The reservoir was officially inaugurated in 1965. The Welsh population ever since has had the feeling that the British government is ready to sacrifice its land to protect and promote English interests, but pays special attention to environmental considerations when it comes to creating jobs in a stricken area. This is clearly illustrated in the following extract taken from the newspaper *The Times* published on 17 February 1958:

> Unhappily, this situation, coming on top of the deep resentments aroused in Wales by Liverpool's Tryweryn reservoir project, has aroused an emotional bitterness of which those who can read the Welsh language weekly papers can have no doubt. The argument is that the English think themselves free to dispossess Welsh farmers in order to build reservoirs for English needs, but that when there is a project which may give unemployed Welshmen some work innumerable societies (almost all of them English) arise to defend the beauty of the scene (*The Times* 1958: 9).

When devolution was introduced in Wales, many journalists considered that these events, happening at the time of the Parliament for Wales Campaign, allowed to arouse the Welsh spirit, as Patrick Fletcher who explained in an article published in the Welsh daily *The Western Mail* on 6 May 2002 and entitled "Wales is at the Heart of the Argument over Water" that "Tryweryn helped make Welsh language and culture a political issue. [...] Wales was powerless to have a say in its own destiny. The opening of the Tryweryn reservoir in October 1965 sparked protests that eventually led to the creation of the Welsh Assembly" (Fletcher 2002). Even Dafydd Wigley, Plaid Cymru's former president, admitted that it was Tryweryn and the events that happened there in his youth that encouraged him to become a politician in order to defend the Welsh assets thanks to an elected assembly. An Assembly eventually set up in 1998 and described in the following terms in *The Legacy Report 2007-2011*: "The National Assembly for Wales is the democratically elected body that represents the interests of Wales and its people, makes laws for Wales and holds the Welsh government to account" (NAW 2011: 4). Environmental stakes thus represent a nationalistic niche in Wales insofar as questions of identity and the protection of the environment are deeply connected. Such a feeling is shared by Rory Francis, the communication manager of ENGO Coed Cadw (Wales Woodland Trust):

> There is a great pride in Wales regarding the quality of our environment, and a determination to protect it. To a certain extent, I would argue that this is linked to the Welsh culture and the Welsh language. I believe this has been an important factor in campaigning to protect and enhance the Welsh environment (interview given in March 2012).

He considers that such an interconnection is one of the reasons explaining the central position of the environment in the devolution settlement in Wales.

Sustainable development is indeed one of the key principles of the Government of Wales Act 1998, a legal document which can be regarded as the founding text of the NAW, and was reiterated in section 79 of the Government of Wales Act 2006 stating that "The Welsh Ministers must make a scheme (« The sustainable development scheme ») setting out how they propose, in the exercise of their functions, to promote sustainable development" (Great Britain 2006: section 79). The Welsh institutions have to respect two requirements: they have first to regularly publish a sustainable development scheme and then to assess the actions led during each assembly. *One Wales, One Planet*, which came out in May 2009, is the latest scheme. As early as the foreword, Rhodri Morgan, then First Minister, very clearly describes the environmental stake:

> I am proud that the Government of Wales Act 2006 places the promotion of sustainable development at the heart of the Welsh Assembly Government's work. We remain one of the few administrations in the world to have such a statutory duty, and it gives us an opportunity to develop Wales as a small, smart nation, in ways which contribute sustainably to people's economic, social and environmental wellbeing (WAG 2009: 4).

Sustainable development must be a core principle ruling the NAW's actions. The Senedd, the Assembly building inaugurated in 2006, is itself a metonymic place since it respects the environmental norms imposed when it was built. It was made as much as possible with Welsh materials and its fanning and heating systems are quite revolutionary. In July 2009, the Senedd became "the greenest governmental building in the United Kingdom", as officially announced on Green Britain Day (10 July), and, three months later, one of the most energy efficient public buildings in the United Kingdom. At the end of the 20th century, the Welsh voice, thanks to the green speech, seemed at last to be heard.

It first seemed to be heard on the British stage since Wales, for the first time in its history, could devise its own programmes and implement its own policy. The government keeps organising public consultations and publishing official documents, thus giving the impression of monopolising the green speech. For instance it published its very successful charter in May 2010 entitled *One Wales: One Planet, Sustainable Development Charter*. It aims to encourage the organisations and companies operating in Wales to make environmentally-friendly long-term decisions. In February 2012, 117 of them, including Marks & Spencer, Pricewaterhouse Coopers, Lloyds TSB and many SMEs, all central in the Welsh economy, publicly committed themselves to put sustainability at the heart of their work. The Charter is also supported by public organisations such as Cardiff University or British Waterways, and the voluntary sector, for instance The National Trust[1] and Keep Wales Tidy[2].

The Welsh voice is also heard on the European stage since Wales would like to be acknowledged in Europe as a green model, which the government underlined in a document published along with the previously mentioned *Sustainable Development Charter*: "Wales is recognised internationally as a leader in sustainable development, and learns from, and exports its learning to other small nations and regions in Europe and wider afield" (WAG 2010: 10). That is why Wales is a member of Nrg4SD, the Network of Regional Governments for Sustainable Development, an organisation set up in the 2002 Johannesburg Earth Summit by a group of regions committed to implement sustainable development policies and gathered in Cardiff in March 2004. The network aims to promote sustainable development at the level of subnational governments worldwide, and encourage exchange, partnerships and collaboration among its members, as well as between them and other major international stakeholders. On 12 March 2005 in Lake Toba, Nrg4SD adopted its strategy for the following years and detailed the major elements necessary to allow regions to commit to sustainable development:

> Regions are considered from the point of view of proximity, efficiency and spatial efficiency, to be strategically located as a necessary and crucial sphere of government for the development of policy for and implementation of sustainable development (Nrg4SD 2005: 6).

Regions were to become major decision-making places to set up a sustainable development policy. The Welsh institutions have met with some success, managing for instance to ban GMOs in the country in March 2004 while the British government was allowing some seeds in England. That was a commitment clearly stated in *Farming for the Future*, a document published by the NAW in November 2001: "Action 52: The

[1] The National Trust was founded in 1895 "with the aim of saving our nation's heritage and open spaces", it is defined in the following way on the website of the organisation: "We're a UK conservation charity, protecting historic places and green spaces, and opening them up forever, for everyone", www.nationaltrust.org.uk, accessed in July 2012.

[2] The Keep Wales Tidy Campaign was set up in 1972 with Welsh Office funding and, in 2005, Keep Wales Tidy was created as an independent organisation working towards achieving a clean, safe and tidy Wales after a devolution process. See www.keepwalestidy.org, accessed in July 2012.

National Assembly Government will continue to operate the most restrictive policy possible in respect of Genetically-Modified Organisms, within the framework of European legislation" (NAW 2001: 36). And yet, turning words into concrete actions rapidly turned out to be difficult for the new institutions.

From words to actions

The NAW has been faced with many obstacles, especially institutional. It long suffered from the fact it only had secondary legislative powers, contrary to Scotland, where the Parliament has had primary legislative powers ever since 1998. Such a difference, often justified by the differing history of the two countries, has been a source of constant dissatisfaction in Wales since the devolution process. The NAW's power of decision in the devolved matters was sharply limited. Its prerogatives evolved in 2011 following the Yes vote in the March Referendum on the NAW's law-making powers. The Assembly Members (AMs) elected in May are able to vote Welsh laws, and as early as 12 July 2011 First Minister Carwyn Jones introduced in the NAW the legislative programme of the new government, a set of measures published a few days later. Two major bills were mentioned: the Sustainable Development Bill[3] allowing to include sustainable development in all the actions not only of the NAW, but also of the public bodies, and the Environment Bill, aiming to manage the environment in an integrated way in order to enhance environmental protection and deliver ecological gains, while easing the regulatory burden. The government thus takes into account the environmental stakes, so important for Welsh people who, for 91% of them according to the previously mentioned poll, support the idea that caring for the environment is a key element of living in Wales. It is currently writing out the first bill, which is welcomed by Jonathon Porritt, the Chair of the UK Sustainable Development Commission between 2000 and 2009, who wrote in an article published in *The Guardian* on 11 July 2012 and entitled "Wales Must Set an Example with its Sustainable Development Bill":

---

[3] In his speech to the AMs, Carwyn Jones described the government's objectives in the following way: "We will legislate to embed sustainable development as the central organising principle in all our actions across government and all public bodies, by bringing forward a sustainable development Bill. The approach will set Wales apart as a sustainable nation, leading from the front", The Welsh Government's Legislative Programme 2011-2016, www.yoursenedd.com, accessed in July 2012.

> Despite the gloomy failures of Rio+20 and David Cameron's empty "greenest government ever" promise, it's good that some parts of the UK can still come up with an enlightened approach. At the moment, it's Wales that provides a bright light in these gloomy times. In Cardiff, the devolved administration, led by first minister, Carwyn Jones, is clearly still committed to the idea of sustainable government. The Welsh assembly is taking a very different approach to the UK coalition. It's legislating to put sustainable development at the heart of all government and public sector decisions, as well as continuing the legacy of the Sustainable Development Commission (SDC), that was foolishly scrapped by Westminster (Porritt 2012).

It seems that the green speech is now clearly delivered by Wales.

There is nevertheless a bone of contention between Cardiff and London: energy, which is a partially devolved matter since the Welsh institutions can only manage power schemes generating under 50 MW. That is really problematic when dealing with the building of nuclear stations. Even if Wales is and has always been fiercely opposed to nuclear energy, and thus to any new nuclear build in the country, David Cameron's government announced in June 2011 it had decided to build new generation reactors in Wylfa, where is located the country's second station supposed to stop any activity more than ten years before. After the shock and outcry caused by the decision, there were strong reactions in Wales. In a document entitled *Energy Wales: A Low Carbon Transition* and published on 14 March 2012, First Minister Carwyn Jones and Environment Minister John Griffiths supported the development of nuclear energy on Anglesey, for the very first time in the Welsh government's history:

> The development of the Horizon nuclear new build (Wylfa B) is a vital component of not just the Anglesey Energy Island Programme but of our wider energy future in providing a constant energy source to complement the intermittency of renewable sources. [...] The Welsh Government support the development of a new nuclear power station on Anglesey (WAG 2012: 21).

Such a decision, announced only a few days after the anniversary of the nuclear accident in Fukushima, Japan, was sharply criticised by the environment defenders who reminded that in 2008, in its *Renewable Energy Route Map*, the government had committed to promote renewable energy in order to ensure that, within 20 years, Wales would produce more energy than it consumed, and all from renewable sources[4]. They also raised the issue of the NAW's powers, comparing the NAW with the Scottish Parliament. On the publication of *Energy Wales*, Dr Calvin Jones, an expert on environmental energy infrastructure at Cardiff Business School, quoted in an article published by Graham Henry "Wales Losing the Green Race to Scotland because it Lacks Policy Powers, Claim Experts", insisted on the fact that Wales only had limited powers in the energy sector, contrary to Scotland:

> Scotland has more control over its energy policy than Wales does, so even if the ambition was as great as Scotland's, we don't have the levers to enact a lot of it. The low-carbon aspiration in Wales, as Carwyn's document this week has shown, has focused on nuclear being part of the mix, while that is certainly not true of Scotland (Henry 2012).

He was agreed with by Gareth Clubb, chairing FoE Cymru, who, in a press release, was calling for more energy powers for the Welsh government:

> Wales desperately needs the power over planning and consenting for major energy infrastructures to free ourselves from the Whitehall mandarins who have no interest in Wales taking the lead in renewable jobs. Devolution of powers over energy is working for Scotland, and if we're not going to be left behind, we must see it coming to Wales. It's good for the environment, good for job creation, and it's the only way Wales can play a role in reducing our climate change emissions. That is what we need to start hearing from the Welsh Government (Clubb 2012).

---

[4] Environment Minister Jane Davidson, in Renewable Energy Route Map, published on 19 February 2008, explained: "In particular we believe that with Wales' coastline, geography and climate, it is quite feasible for us within 20 years to produce more electricity from renewables than we consume as a nation", p. 1. A commitment reminded by Chris Kelsey, dealing with the Welsh government's decision in an article published in The Western Mail on 6 April 2011 "From nuclear power to offshore wind, difficult choice for energy planners": "This must be a matter of frustration for the men and women in the Senedd. Three years ago the Assembly Government issued a consultation on what it called its renewable energy route map, in which it asserted that within 20 years Wales could generate more electricity than it consumed, and all from renewable sources. Nuclear power did not feature in the route map, not only because the Assembly Government has no power over whether nuclear stations are built or not, but also because to the green-minded Assembly ministers nuclear energy is not a truly sustainable option in the way wind and wave power are".

Considering the traditional opposition to nuclear energy in Wales, how can the government's u-turn be explained? Does it want to give the impression that the decision to keep producing nuclear energy in Wylfa is being made in Cardiff, and not in London? And thus to preserve its authority and credibility? Or is it simply yielding to economic pressure?

It may be difficult to reconcile economic and environmental stakes, which Robert Minhinnick, FoE Cymru's co-founder, asserted as early as 1994 in *Green Agenda – Essay on the Environment of Wales*, a book he directed: "Even devolution of power in Wales poses environmental threats. It is interesting to speculate how a Welsh senedd with its inevitable core of party yes-men and stalwart councillors nominated for seats as testimonials to loyalty, would cope with the blandishments of big business wishing to exploit oil and gas around our coasts; water, coal and other minerals ashore; or further develop Wales as a tourist destination" (Minhinnick 1994: 21). Such a conflict of interests was illustrated with the Severn Barrage, a major scheme in the Welsh government's energy route map dropped by the British government in October 2010 for financial reasons, in the context of a worldwide economic crisis. That was when it decided to study possible sites for new nuclear power stations to be built. The plan was also an example of the discrepancy between the protection of the environment and sustainable development since most environmentalists opposed a scheme sharply disrupting the local ecosystem but supposed to produce renewable energy on a large scale.

As a result, Welsh people felt betrayed by their institutions which seemed to be unable to implement their programmes, to turn words into actions. They thus lost interest in the NAW, as shown by the very low turnout for the different elections organised ever since its creation in 1998: only 45,9% of Welsh voters came to elect their AMs in May 1999, then 38,2% four years later and 43,7% in 2007. Fewer than 36% of voters gave their opinion for the referendum on the Assembly's law-making powers in March 2011. When the Severn Barrage scheme was dropped, Roger Harrabin, an environment analyst for the BBC, listed the three main reasons why the scheme was axed. One of them, in his opinion, was the opposition of the green lobby: "There were three reasons why the barrage was turned down. Firstly, environmentalists are heavily split over it, and pressing ahead with such a controversial scheme without the wholehearted support of the green lobby, which is championing renewable energy, would have been politically risky" (Harrabin 2010). Is the green speech more and more delivered by ENGOs, lobbies which make them heard in Wales, more than politicians themselves?

The green speech in Wales

In the 1970s, ENGOs developed worldwide, especially following the Stockholm Earth Summit convened by the United Nations in 1972 and gathering the governments of many countries. Were thus created WWF, first World Wildlife Fund, then World Wide Fund for Nature, and now simply WWF, in 1961, Friends of the Earth International (FoEI) in 1971, and Greenpeace, which turned international in 1979. These ENGOs quickly decided to open local branches. They have varied fields of action: from energy issues, such as nuclear pollution or the plans to build new nuclear stations, the expansion of opencast mining and the development of wind energy, to the problems related to air pollution, waste management, spatial or landscape planning. They can be highly specialised, as the Royal Society for the Protection of Birds, which nevertheless has 51,000 members just in Wales, or Coed Cadw, mentioned before, which has 8,000. The most famous ENGO is most certainly FoE Cymru, founded in 1984 by Margaret Minhinnick and composed then of 600 members, against 3,000 ten years later, and several thousand today according to Gareth Clubb. It is regularly quoted in the press and has met with several successes: in 2003, the NAW targeted to produce 10% of Welsh electricity through renewables by 2010 following a campaign led by FoE Cymru. Its biggest victory, for its members, remains section 121 of the GOWA 1998, ie the obligation put on the NAW to promote sustainable development in all its decisions. Yet, all these ENGOs worked on their own, in isolation, which Margaret Minhinnick herself regretted: "Most campaigners insist on looking only at their own backyards and are unwilling or unable to link issues, or network meaningfully with other groups" (Minhinnick 1994: 66). That, in her opinion, was the *sine qua non* condition for the success of the green lobby, a phrase commonly used to describe ENGOs.

The numerous Welsh environmental associations realized they were much more powerful when they worked together and collaborated. That was the reason why they decided in 2002 to set up the Wales Environment Link (WEL), from the Wales Wildlife and Countryside Link created 12 years before. WEL now numbers 27 member organisations, including the Council for the Protection of Rural Wales (CPRW) or WWF Cymru, accounting themselves for 245,000 members in Wales, out of a population of nearly 3m people. In other words, even if the same person can belong to several associations, it still means that around 8% of the Welsh population is involved in WEL. It works as an ENGO network, and represents the official link between the government and the environmental voluntary sector. This organisation is the mouthpiece of the numerous Welsh ENGOs, which it states itself in a document published in November 2011 *In Touch with Environmental Non-Governmental Organisations in Wales*: "WEL is an advocate of the memberships' collective views and a voice for the environmental movement in Wales" (WEL 2011: 4). As a consequence, the barrier separating the political and voluntary communities is getting blurred. ENGOs tend to look more and more like political parties: their leaders are generally well-known, even more than the leader of the Wales Green Party; they enjoy a wide media coverage and are frequently quoted in the press insofar as their press releases are widely publicized; and, last but not least, they do not hesitate to publish a manifesto for each election campaign in Wales. As WEL which published *Wales Environment Link, Manifesto, Policy Priorities for Wales 2011-2015*, a 15-page document written in September 2010, the objective of which was clearly defined in the introduction: "In this manifesto, Wales Environment Link has set out the steps that we consider the future Welsh Assembly Government should take in order to ensure a sustainable future for the people and environment of Wales" (WEL 2010: 2). The very authors thus consider this document as a "manifesto". It is divided into 12 parts dealing with the following subjects: access and recreation, agriculture and food, biodiversity, climate change, energy, forestry and woodland, historic environment, marine, planning, recycling, transport and water. These titles remind us of the manifesto published by the Wales Green Party, *The Green Deal*. A double page is dedicated to each subject, with a similar layout: the priorities for Wales are first clearly listed, then the context is given, with many photos included to make it nicer and more pleasant. This is thus a very pedagogical and reader-friendly document, which really looks like a political manifesto. Just as the one published by FoE Cymru on 11 March 2011, ie only a few days after the referendum on the Assembly's law-making powers, and entitled *Manifesto for a Greener Wales*, with the following subtitle "A realistic programme for the next Assembly Government". To finish, the careers made by famous members of the green lobby reveal the links between the two communities. Jonathon Porritt, for instance, directed FoE England from 1984 to 1990, and then became the administrator of WWF Cymru until 2005, after co-leading the UK Ecology Party. In 1996, along with Sara Parkin and Paul Ekins, he founded Forum for the Future, an independent ENGO publishing support and advice documents aimed at economic and political decision-makers. Between 2000 and 2009, he chaired the UK Sustainable Development Commission. Are then the two communities, the political one and the green lobby, fighting for power?

Gareth Clubb, dealing with the power held by ENGOs in Wales, indicated in an interview given to the author in March 2012 that "They are influential in shifting the environmental debate. Without them, the political parties would have little political space in which to promote environmental messages". They are thus supposed to work in collaboration with the Welsh institutions in order to define the green agenda. And yet, they rather seem to be more often than not in conflict with them. When *One Wales: One Planet* was officially launched in May 2009, Rhodri Morgan, then the WAG's First Minister, insisted on his will to deal with environmental stakes, especially sustainable development, in order to strengthen the credibility of the NAW and the WAG, accused of being concerned only with secondary issues and of meeting with little success for major issues:

> We must build sustainability into everything we do. This is not just an issue for the green lobby. It is not just about recycling our cans. It is about ensuring Wales can sustain itself with sustainable jobs and a sustainable quality of life. [...] There is no room for short-term decisions (Morgan 2009).

A part of the green lobby nevertheless won a victory in 2011 when the government announced it intended to limit the building of new onshore wind farms, which was in contradiction with its previous commitments. In *Renewable Energy Route Map for Wales*, already mentioned, wind energy was at the heart of the NAW's energy policy: "Wind energy is the most readily available commercial renewable technology and Wales' weather and geography means we are well-placed to use it" (WAG 2008: 23). The government went on promising to keep developing this type of energy: "We will continue to pursue the proposals in TAN 8 and monitor the uptake of wind farm sites before undertaking a further review in the light of this and related consultations" (WAG 2008: 26). And yet, three years later, it backtracked, as underlined by Chris Kelsey in an article published on 17 June 2011 and entitled "Carwyn Hands the Anti-wind Lobby a Victory":

> The Welsh Government appears to have bowed to the angry voices of anti-wind protestors and is seeking to limit the development of wind farms in Mid Wales. [...] He is careful to preface all this with a nod to Wales' obligation to meet carbon reduction and renewable energy targets. But the fact is the Welsh Government has handed a victory to the anti-wind lobby (Kelsey 2011).

The journalist considers that the First Minister gives hope back to environmentalists opposing biomass. In this case, ENGOs shouted loud enough! It must be noted that statistics indicate winds in the British Isles weaken and the year 2010 is said to have been the quietest year of the last decade (an average 20% decrease in wind speed compared with 2008).

To conclude, environmental stakes are at the heart of the devolution settlement in Wales and the Welsh government uses them to promote the identity and the specificity of the country, both inside the United Kingdom and in Europe. Nevertheless, the NAW, faced with many obstacles, especially institutional and economic ones, has kept disappointing the Welsh, anxious to protect their environment. Turning words into actions is no easy task, as experienced by the various Environment Ministers. So much so that the Welsh ENGOs, working sometimes as political parties and better and better organised, have taken over the green speech and put pressure on the Welsh institutions. They are meeting with more and more success. Is the situation similar in England? Are English ENGOs playing the same role? The green lobby seems to be weakening in Wales' neighbouring country. Reacting to a speech made by George Osborne, David Cameron's Chancellor of the Exchequer, many defenders of the environment, as Wildlife Trusts which has more than 800,000 members, decided to send letters to *The Observer* in order to denounce the anti-green policy carried out by the Coalition Government. Such a move was welcomed by *The Guardian*: "An extraordinary alliance of countryside campaigners, wildlife groups and green activists has launched a savage onslaught on the government, accusing it of showing "stunning disregard" for the environment" (McKie&Helm 2011). George Monbiot, Jonathon Porritt and Tony Juniper, both of them former chairs of FoE, and Caroline Lucas, the only green MP and the leader of the Green Party of England and Wales, wrote in an open letter addressed to *The Observer* and quoted in *The Guardian* that "Osborne has proclaimed that protecting the environment is against the public interest – something no senior politician in this country has done in recent history". The day after the letter was published, a Welsh blog called Inside Out reacted with the headline "Is the Green Lobby a Spent Force?" The author commented the words used in the mentioned letter in the following way: "Yet there is otherwise little press interest"[5], as if the English green lobby had become dumb.

[5] Inside Out, http://insideoutswansea.blogspot.com, accessed in January 2012.

Bibliography

Anon, 'Trawsfynydd', *The Times*, 17-02-1958.

Clubb, Gareth, 'Welsh government's energy vision just tinkering around the edges', FoE Cymru's press release, 14-03-2012, www.foe.co.uk, accessed in April 2012.

Fletcher, Patrick, 'Wales is at the Heart of the Argument over Water', *The Western Mail*, 06-05-2002.

Great Britain, GOWA 2006, Elizabeth II (Chapter 32), London: HMSO, 2006.

Harrabin, Roger, analysing an article 'UK Government Axes Severn Estuary Barrage Plan', 18-10-2010, www.bbc.co.uk, accessed in July 2012.

Henry, Graham, 'Wales losing the green race to Scotland because it lacks policy powers, claim experts', 17-03-2012, www.walesonline.co.uk, accessed in April 2012.

McKie, Robin & Helm, Toby, 'New green alliance in savage attack on George Osborne', *The Guardian*, 03-12-2011.

Minhinnick, Robert, *Green Agenda – Essay on the environment of Wales*, Bridgend: Seren, 1994.

Morgan, Rhodri in a speech delivered during the Hay Festival, when *One Wales: One Planet* was published, in « 'Zero Waste' Wales plan unveiled », BBC, 22-05-2009, www.bbc.co.uk, accessed in April 2012.

National Assembly for Wales, *Farming for the Future*, Cardiff: NAW, November 2001.

—. *The Legacy Report 2007-2011*, Cardiff: NAW, 2011.

Nrg4SD, *Strategy 2005-2011*, 12-03-2005, www.nrg4sd.net, accessed May 2008.

Porritt, Jonathan, 'Wales Must Set an Example with its Sustainable Development Bill', *The Guardian*, 11-07-2012.

Wales Environment Link, *Wales Environment Link, Manifesto, Policy Priorities for Wales 2011-2015*, *WEL*: Cardiff, 2010.

—. *In Touch with Environmental Non-Governmental Organisations in Wales*, *WEL*: Cardiff, November 2011.

Welsh Assembly Government, *Renewable Energy Route Map for Wales*, *WAG*: Cardiff, 2008.

—. *One Wales: One Planet*, Cardiff: WAG, 22-05-2009.

—. *Guidance and Advice to support the SD Charter*, Cardiff: WAG, March 2010.

—. *Energy Wales: A Low Carbon Transition*, Cardiff: WAG, 14-03-2012.

THE SNP'S ENVIRONMENTAL POLICIES: A VISIONARY ECONOMIC PROJECT FOR AN INDEPENDENT SCOTLAND OR A "GREEN" UTOPIA?

FIONA SIMPKINS

UNIVERSITY OF LYON II

Environmental and energy concerns are increasingly considered as some of the most salient issues faced by Western nations today, confronted not only with the impact of global warming but also with fast-diminishing world oil reserves and a subsequent growing dependence on some Middle Eastern countries. This is the case of the UK as further North Sea oil extraction is now estimated to a mere thirty years. In fact, a recent study from the University of Aberdeen[1] warned that although North Sea oil and gas production could rise over the next few years to up to 1.4 million barrels per day in 2017, as large new fields are being developed, this will not be able to reverse a continuous decline since 1999. Reserves are now likely to amount to 16.8 billion barrels of oil equivalent (boe) over the next three decades, which is less than the Department of Energy and Climate Change's previous estimate of 20 billion barrels of oil equivalent. This situation is all the more worrying in a climate of economic recession, and the cheaper nuclear option – on the French model for instance – appears increasingly appealing to successive British governments. Thus in January 2008, the Brown Labour government officially authorized the construction of new nuclear power stations as well as the modernisation of existing British nuclear power stations. The plan was largely followed by the current coalition government elected in 2010, even after it faced strong anti-nuclear pressure in the aftermath of the Fukushima disaster in Japan: it is planned that eight new nuclear power stations should be constructed

---

[1] Professor Alex Kemp & Linda Stephen, *Prospects for Activity in the UK Continental Shelf after Recent Tax Changes: the 2012 Perspective*, University of Aberdeen, December 2012.

on pre-existing sites in Britain. Indeed, according to the Carbon Plan[2] published by the Department of Energy and Climate Change in December 2011, the current coalition government considers that the nuclear option constitutes the cheapest solution today in order to reduce carbon dioxide emissions and their subsequent impact on the environment whilst protecting the UK from a certain economic vulnerability to highly volatile oil and gas markets.

However, although similar objectives have been adopted by the Scottish Government in Edinburgh, the solutions it wishes to provide in order to meet them diverge considerably. Indeed, the SNP, in power since May 2007, is traditionally opposed to nuclear energy and has systematically referred to the Scottish Parliament's planning powers in order to prevent any nuclear development on Scottish soil. On the contrary, the SNP Government is seeking to make of Scotland a world leader in terms of renewable energy production and technology, thus ensuring an independent Scottish energy production. Opposition to nuclear power and a strong commitment to the development of renewable energy in Scotland were at the heart of the SNP's coalition agreement with the Scottish Green Party in May 2007 (Scottish National Party and Scottish Green Party Cooperation Agreement), but were further developed after the SNP succeeded in winning the May 2011 Scottish Parliament elections and forming the first Scottish majority government. A "greener Scotland" was indeed one of the SNP's headline commitments in its 2011 electoral manifesto and it is worth questioning the extent to which "green" policies can serve the SNP's main political objective of an independent Scotland? The association of an independent Scottish energy production with the SNP's *raison d'être* - a politically independent Scotland - raises the question of the SNP's economic strategy for an independent Scotland and the weight that the latter shall carry in the 2014 independence referendum.

The impact of environmental issues in Scotland

Environmental issues are often considered to constitute a new political field that emerged during the 1980s and 1990s. Yet, not only are environmental policies in the UK rooted in much older policies seeking to improve air and water quality for instance, but the environment itself can hardly be seen as a political field in itself as its boundaries are permeable with those of other areas such as economics, transport or agriculture.

[2] *The Carbon Plan: Delivering Our Low Carbon Future,* Department of Energy and Climate Change, HM Government, December 2011.

Furthermore, if environmental policies transcend administrative and political boundaries, they also escape any formal definition in that they are the result of a wide variety of distinct cultural and historic mental constructs. Indeed, a March 2008 European study[3] found that 31% of people polled in the UK associated the environment with city pollution against 19% only in France for instance. In Poland and Cyprus, 26% and 31% of respondents respectively associated the environment with beautiful landscapes, whilst 21% of Czech respondents associated it with the protection of nature.

The environment has nevertheless been considered as a consensual issue or valence issue for the last few years as it meets with very little opposition. But how much political weight does it carry with the public? In fact, the results of the Scottish Environmental Attitudes and Behaviours Survey[4] published in 2009 suggest that the environment is not perceived as a political priority in Scotland. Indeed, 38% of respondents mentioned the economy, 32% crime, law and order, and anti-social behaviour, 17% the Scottish constitution and 12% only the environment and climate change. It is noteworthy, however, that the latter figure is slightly larger than that found by a similar UK poll during the same period (8%)[5]. Furthermore, many more Scottish respondents considered the environment as a global priority (35%), second only to terrorism (43%). Finally, the study results on Scottish attitudes to environmental issues stressed that a majority of Scots (41%) believed that climate change was the most important environmental issue today, well before city pollution, which the aforementioned European study had placed as the top priority of overall British respondents. However, few believe that improving the environment should be a priority for the Scottish Government: only 2% of respondents believed that this should be the case according to the Scottish Social Attitudes Survey 2011 whilst a majority (36%) believed that improving the economy should be the Scottish Government's top priority as shown in the table below.

---

[3] "Attitudes of European citizens towards the environment", Report, *Special Eurobarometer* 295 Wave 68.2, *TNS Opinion & Social*, mars 2008, p.6

[4] *Scottish Environmental Attitudes and Behaviours Survey*, Scottish Government Social Research, March 2009.

[5] Annual average of the Ipsos MORI poll series "Most important issues facing Britain" (www.ipsos-mori.com/content/the-most-important-issues-facing-britain-today.ashx).

Table 1: What should be the Scottish Government's highest priority? (2004-2011)

%	2004	2005	2006	2007	2009	2010	2011
Help the economy grow faster	16	18	14	16	33	37	36
Cut crime	22	23	24	27	17	18	17
Improve people's health	27	26	24	17	13	15	16
Improve standards of education	17	15	16	12	15	12	15
Improve housing	12	10	12	16	13	9	11
Improve the environment	4	4	6	5	3	3	2
Improve public transport	1	1	2	1	1	1	-

Source: Scottish Social Attitudes Surveys 2004-2011

The ways in which the environment is perceived, however, and the political weight environmental issues carry with the public vary from one nation to the next within the UK itself, as suggested by the aforementioned poll results. In so far as environmental perceptions correspond to distinct cultural and historic constructs, they are inevitably associated to the national identity of each nation. Interestingly, the 1997 Scottish and Welsh Referendum studies included questions on national pride and cultural symbols and found that a vast majority of Scottish and Welsh respondents (84% and 72% respectively) mentioned landscapes as their greatest source of national pride, which is one aspect of the environment at large.

But to what extent do the SNP's environmental and anti-nuclear policies resonate with the Scottish public? In fact, a YouGov/Express poll of January 2008 only found 21% of respondents in favour of a complete pull-out from nuclear energy if Scotland were to be independent, against 71% of respondents in favour of the continued use of diverse sources of energy, including wind power, coal, gas and nuclear energy. This trend was confirmed by an April 2012 poll for the Sunday Herald that found that 39% of respondents strongly agreed and a further 33% tended to agree with a statement supporting "continuing development of wind power as part of a mix of renewable and conventional forms of electricity generation". On the other hand, the results of the Scottish Social Attitudes Survey of 2007, which included a question on the level of satisfaction regarding the SNP's decision to oppose any new nuclear station construction project, found 51% of respondents favourable to the SNP's anti-nuclear policy. This may translate a stronger satisfaction with the fact that the SNP were opposing a Westminster governmental decision rather than true anti-nuclear sentiment in Scotland however.

It is no wonder then, that the SNP should be particularly keen to develop environmental policies and present itself as the champion of a "greener Scotland". Although the environment is not considered as a political priority by a majority of Scots, it nevertheless remains a consensual issue which, as we have seen, remains in some aspects narrowly linked to the Scottish national identity and sentiment. In defending the idea of a "greener Scotland", the SNP can thus be seen as defending Scotland's interests. This is particularly important in the context of Scottish Parliament elections – as opposed to Westminster elections – as electoral trends since 1999 show that the Scottish electorate tends to be more likely to vote SNP because it is seen as a Scottish party devoted to defending Scottish interests only rather than UK-wide parties such as Labour, the Conservative or the Liberal-Democrat party for instance.

Green policies in the separatist discourse

As suggested earlier, the environment is not limited to a distinct political field but is often included in other fields, especially that of economics. Indeed, the traditional late twentieth century debate between environmental policies and economics – with the former protecting the environment but failing to develop economic growth and the latter expanding the economy but destroying the environment – was ended with the notion of sustainable development and the idea that the environment and economics can no longer be dissociated. The integration of environmental concerns to the economic policies of modern states is widely accepted today. It nevertheless presents further electoral advantages for the SNP in that the latter is able to fully associate its economic programme to a separatist discourse. Indeed, by integrating a powerful valence issue such as the environment to their economic programme, the nationalists are able to widen the appeal of their economic policies and convincingly stress the inadequacies of the current constitutional set-up for Scotland.

Although the Scottish Parliament has important powers in terms of environmental policies, the permeability of that field's boundaries often mean that the Scottish Government has to abandon certain projects because of the cross references between environmental issues and matters reserved to Westminster. The Scottish Parliament has large legislative competences in terms of transport, planning, heritage and the environment. The latter include environmental protection, matters relating to air, land and water pollution and the functions of the Scottish Environment Protection Agency; water supplies and sewerage; and policies designed to promote sustainable development within the international commitments agreed by the UK. But in fact, the latter represent a sticking point as the majority of international agreements are signed by British representation groups that do not necessarily include representatives from Scotland. Scottish representatives are sometimes invited to join the British representation groups as advisors or observers but Scotland cannot send its own representation groups to defend its own opinions, policies or interests outside of any UK-wide concerns.

This issue becomes particularly sensitive when European environmental policies are concerned. Indeed, since European legislation takes precedence over British – and Scottish –legislation, Scottish policies in terms of fishing and fish farming, marine conservation, agriculture, forestry, rural development, animal health, biodiversity, waste and recycling, drinking water, pollution and climate change, for instance, are determined for the most part by decisions made in Brussels. Although section 53(2)(a) of the Scotland Act 1998 effectively transferred the responsibility concerning the application of European policies to the Scottish devolved administration, foreign affairs – and therefore the right to participate in the European legislative process – remain part of Westminster's remit. Similarly, if the cooperation principles underlying the relations between the British and Scottish governments are clearly stated in the Memorandum of Understanding and Concordats, the latter have no legally binding value and, in the end, the decision to invite a Scottish Government representative to observe or speak at a European Council meeting belongs entirely to the British minister concerned. Indeed, the viewpoint presented by the United Kingdom Permanent Representation (UKREP) at European Council meetings must be representative of the United Kingdom as a whole and of the British Government. So even though concordats include certain mechanisms allowing for the preliminary contribution of Scottish ministers to the British policy adopted, the interests that the latter defend often fall victim to negotiations with other European states on the basis of priorities that do not necessarily correspond to those of Scotland. These drawbacks, linked to the very nature of the new British post-devolution constitutional setup, are thus consistently denounced by the Scottish nationalists in order to fuel the independence debate in Scotland, and contribute to some extent to the importance of the environment in the current SNP Scottish Government discourse as some of the following examples indicate.

Fishing and fish farming are often used by the SNP Government to stress the inadequacies of devolution for instance. Scotland is one of the most important sea fishing nations in Europe, it represents approximately

66% of the total UK volume of fish and is the largest producer of farmed Atlantic salmon. Facts that the SNP Government will stress as it repeatedly demands that Scotland should be allowed to lead any European negotiations concerning fishing and fish farming on behalf of the UK, denouncing the unfairness of the current system that "remains an impediment to the successful achievement of wider goals"[6]. In fact, the European Union has adopted certain mechanisms allowing a region, such as Flanders in Belgium for example, to represent a member-state in such circumstances.

Another environmental sticking point between Westminster and Edinburgh which the SNP has integrated to its separatist discourse concerns the European Common Agricultural Policy. Indeed, both governments diverge fundamentally over the CAP as the British Government wishes to phase out the First Pillar of the CAP, regarding income support for farmers delivered through the Single Farm Payment as well as actions to intervene in markets for certain products, in favour of the development of the Second Pillar, consisting of rural development programmes. The Scottish Government on the other hand is opposed to its British counterpart's reform project of the First Pillar as it fears the negative impact it would have on the Scottish agricultural economy. Indeed, 85% of Scottish soil falls under the EU's "less favoured" agricultural land category with a lower profitability than better quality land and therefore liable to be selected for European aid programmes included in the First Pillar of the CAP. It is worth stressing that the proportion of "less favoured" land is of 80% in Wales, 40% in Northern Ireland and only 17% in England[7], thus inevitably creating conflicting interests between the UK nations over European agricultural policies.

Finally, the lack of Scottish representation in Brussels is also criticized by the Scottish Government when environmental policies voted in Holyrood fall victim to British negotiations in Brussels and UK-wide priorities. This was the case of the reforestation project developed in 2009 by the Scottish Government that set an ambitious target of 25% woodland cover in Scotland by 2050. This project was developed as part of the Scottish Government's policies on cuts in carbon dioxide emissions, rural

---

[6] Scottish Government, Rural Affairs, Environment and Climate Change, *Taking Forward Our National Conversation, Choosing Scotland's Future*, 2009, p.11.

[7] Department of Agriculture and Rural Development, *Review of Support Arrangements for Less Favoured Areas in Northern Ireland,* Consultation Report, Chapter 3, mars 2009 (http://www.dardni.gov.uk/consultations-full-report).

development and forestry development[8]. The Scottish Government's woodland project meant to be much more ambitious that its British counterpart's but was severely limited by the rural development programme adopted by the European Commission for the period 2007-2013 which limits the budget lines allotted to reforestation in Europe[9].

However, European policies do not represent the only constraints imposed on Scottish environmental policies: matters reserved to Westminster under the Scotland Act 1998 severely limit the Scottish political arena's field of action in terms of environmental policies as the SNP Government never fails to point out in order to advance the case for an independent Scotland, or at least the devolution of further powers to the Scottish Parliament. Indeed, many environmental policies voted in Holyrood are limited by the fragmented nature of legislative responsibilities in the UK today. Several examples of the conflicts of interests regarding environmental policies and caused by reserved matters have thus been raised by the SNP in its political discourse in recent years to stress the inadequacies of the current constitutional setup. These include marine and coastal issues: fishing is a devolved matter out to 200 nautical miles but oil, gas and shipping regulations are part of Westminster's remit, even in coastal areas. The Crown Estate is responsible for the use of the seabed whilst coastguard services are run by the Maritime and Coastguard Agency, an executive agency of the British Department for Transport. The situation is further complicated in other sectors within this area as the Scottish Parliament only has executive powers, rather than legislative, over marine renewables but can legislate on marine nature conservation and installations at sea but only out to 12 nautical miles. Furthermore, the reserved nature of marine pollution means that Scotland cannot legislate against litter arising from ships de-ballasting or emptying their fuel tanks to effectively protect the seabeds in Scottish waters. The same applies to ship to ship transfer of oil, which represents a potential threat to the marine environment, as shipping is a matter reserved to Westminster.

In repeatedly stressing the numerous constraints and contradictions imposed by the current constitutional setup in Scotland in the context of a valence issue as powerful as the environment, the SNP Government is also

---

8 Forestry Commission Scotland, *The Scottish Government Rationale for Woodland Expansion*, Scottish Government, 2009.

9 Conseil de l'Union Européenne, Règlement (CE) no 1698/2005 du Conseil du 20 septembre 2005 concernant le soutien au développement rural par le Fonds européen agricole pour le développement rural, *Journal Officiel de l'Union Européenne*, 21st October 2005,
http://ec.europa.eu/agriculture/rurdev/index_en.htm

taking advantage of the strong position offered by the Scottish Parliament to fuel the constitutional debate. Indeed, Scottish Social Attitudes Surveys have consistently shown over the years that Scottish voters trust the Scottish government more than its British counterpart to work in Scotland's long-term interest and that a majority wish Holyrood to have more influence on the way Scotland is run than Westminster. However, only 34% of the population considered that the Scottish Parliament actually had most influence over the way Scotland was run in 2012 against 41% who still believed that Westminster retained the most influence ten years after the referendum. Although the number of people who consider that Westminster remains the most influent political force in Scotland has dropped from 66% in 2001 to 41% in 2012, a majority of the Scottish public still seems to perceive a certain pervasiveness of wider British political interests in Scottish politics. This trend is particularly salient if we are to compare these figures to the percentage of people in Scotland who believe the Scottish Parliament or government ought to have most influence over the way Scotland is run (73% in 2011 and 63% in 2012) rather than the British government in Westminster (13% in 2011 and 24% in 2012) as shown in the tables below. Although the difference in figures is less marked for 2012 than in recent years, it is worth stressing that the gap has remained important in recent years – even after having experienced devolution in Scotland for more than ten years – and the 2012 results have most likely been influenced by the current independence referendum debate.

Table 2: Which of the following do you think has most influence over the way Scotland is run?

Which of these has most influence over the way Scotland is run:	1999	2000	2001	2003	2004	2005	2006	2007	2009	2010	2011	2012
The Scottish Parliament	41%	13%	15%	17%	19%	23%	24%	28%	33%	37%	38%	34%
The UK Government at Westminster	39%	66%	66%	64%	48%	47%	38%	47%	39%	45%	38%	41%
Local Councils in Scotland	8%	10%	9%	7%	20%	15%	18%	8%	11%	7%	13%	10%
The European Union	5%	4%	7%	5%	6%	8%	11%	9%	10%	7%	7%	9%

Source: Results compiled after Scottish Social Attitudes Surveys 2001-2012.

Table 3: Which of the following do you think ought to have most influence over the way Scotland is run?

Which of these should have most influence over the way Scotland is run:	1999	2000	2001	2003	2004	2005	2006	2007	2009	2010	2011	2012
The Scottish Parliament	74%	72%	74%	66%	67%	67%	64%	71%	72%	74%	73%	63%
The UK Government at Westminster	13%	13%	14%	20%	12%	13%	11%	14%	13%	16%	13%	24%
Local Councils in Scotland	8%	10%	8%	9%	17%	15%	19%	9%	11%	6%	11%	8%
The European Union	1%	1%	1%	1%	1%	1%	1%	1%	1%	1%	1%	1%

Source: Results compiled after Scottish Social Attitudes Surveys 2001-2012

The stark contrast between the two polls, however, suggests that the idea of a democratic deficit, that became widespread under the Thatcher governments and spurred the return of devolution on the political agenda in the nineties, remains an underlying current in Scottish politics today. The use of an environmental discourse therefore enables the SNP to keep this idea alive in Scotland, even within the context of devolution, and present the current constitutional setup as insufficient to protect Scotland from the potential harm that British policies may have on Scotland's environment, incidentally one of the greatest sources of Scottish national pride, as we mentioned earlier. The rejection of wider British interests in the way Scotland is run in Holyrood – and furthermore regarding Scotland's environmental interests – creates a favourable context for the nationalists: as the only party in Scotland with no UK counterpart, it is able to present itself as the party best suited to defend Scotland's interests and to feed the constitutional debate in Scotland. Indeed, the SNP's arguments regarding a Scottish representation in Brussels for instance seem to resonate with the Scottish public as the results from the latest Scottish Social Attitudes Surveys confirm that a majority of respondents believed that Scotland would have a stronger voice in the world if it were to be independent (42% if the 2012 results are collapsed).

Table 4: As a result of independence, would Scotland have a stronger voice in the world, a weaker voice, or would it make no difference?

%	2011	2012
A lot stronger	22	11
A little stronger	29	31
No difference	27	32
A little weaker	12	12
A lot weaker	7	10
Don't know	3	5

Source: Scottish Social Attitudes Survey 2012, ScotCen.

Furthermore, the long-lasting idea of a democratic deficit and of negative interference from Westminster in Scottish affairs appears to favour the current Scottish Government in its efforts to convince the Scottish electorate of its good economic record and the future economic prospects of an independent Scotland. It appears that Westminster is most likely to be blamed by the Scottish public when it is believed that standards have fallen and the Scottish Government is most likely to be credited when they have risen as suggested in table 5 below. Indeed, 58% of respondents blamed the British Government in Westminster among those who believed that living standards had fallen in 2011, compared to 6% only who blamed the Scottish Government. This is particularly marked in relation to the economy, where the percentage of respondents who were positive about economic performance and credited this to the Scottish Government increased by 30% between 2010 and 2011 (from 39% to 69%) whilst, on the other hand, more people blamed the UK Government when they believed the economy got weaker (44% in 2010 to 54% in 2011), even in the midst of a world recession!

Table 5: "Credit" and "blame" for standards in public services, the economy and standard of living in the past 12 months (2011)

	…increased (credit)		… fallen (blame)	
% of those who say standards have…	2010	2011	2010	2011
Standards in the NHS are mainly the result of…				
Scottish Government policies	49	58	18	12
UK Government policies	30	15	49	54
Some other reason	16	20	21	21
Standards in education are mainly the result of…				
Scottish Government policies	56	65	30	23
UK Government policies	23	12	40	46
Some other reason	14	14	22	22
Standards in public transport are mainly the result of…				
Scottish Government policies	57	60	37	31
UK Government policies	11	13	12	22
Some other reason	24	19	41	38
Strength of the economy is mainly the result of…				
Scottish Government policies	39	69	14	11
UK Government policies	21	15	44	54
Some other reason	32	10	35	26
General standards of living are mainly the result of…				
Scottish Government policies	37	44	8	6
UK Government policies	33	31	49	58
Some other reason	22	12	36	27

Source: Scottish Social Attitudes Survey 2011, Core Module – Attitudes to Government, the economy and public services in Scotland, Scottish Government, ScotCen Social Research.

This is clearly an advantage for the Scottish Government as the strength of its economic policies appears today as a determining factor for the referendum it plans to organise on Scottish independence in 2014. We shall see that the environment plays a key role in the economic strategy developed by the SNP for an independent Scotland but that this may present the SNP with more problems than it solves insofar as its programme is not immune from frequent contradictions.

The environment as economic engine of an independent Scotland

In fact, the SNP's traditional anti-nuclear stance was transformed into an ambitious "green" economic programme as early as 2007, when it developed several major policy lines aiming at reducing carbon emissions and developing renewable energy sources in Scotland in its electoral manifesto for the May 2007 Scottish Parliament elections. The party assured that it would make cutting carbon dioxide emission levels one of its main priorities if it were to be elected at the head of the Scottish Parliament and foresaw the introduction of an ambitious Climate Change Bill which aimed at reducing carbon dioxide emissions by 3% a year with a long-term objective of 80%, that is 60% more than the target set by the UK Government after the Stern Report was published in October 2006. The alternative to the development of nuclear energy planned by the UK Government would be provided by renewable or "green" energy sources and represent what the authors of the manifesto considered to be "a Scottish energy revolution"[1]. The SNP furthermore envisaged that Scotland should take the lead in world research on the development of renewable energy sources and foresaw the creation of a £10 million international prize, the Saltire Prize, aiming at encouraging the researchers to settle in Scotland and develop new projects on wave power and tidal power. Two prizes have already been awarded at the time of writing: the first to Professor Stephen Salter in 2011 and the second to Richard Yemm in March 2012 for the development of technologies enabling electricity generation through wave power. This serves as concrete evidence of the SNP Government's objective to make of Scotland a leader in renewable energy generation, a recurrent theme in SNP economic discourse as the title of the Scottish Government's 2008 consultation paper on renewable energy indicates: "*Making Scotland a Leader in Green Energy – Draft*

[10] Scottish National Party, *It's Time To Move Forward,* Scottish National Party, Edinburgh, 2007, p. 29

Framework for the Development and Deployment of Renewables in Scotland".

Following the Climate Change Act of June 2009, it appeared that Scotland was indeed on its way to becoming one of the world leaders in terms of renewable energy production. Scotland exports 20% of its electricity production to the rest of the UK today and, in this sense, benefits from a much more comfortable position than its British partners if it is to explore new sources of energy production. Furthermore, the potential for renewable energy production that is offered by Scotland's natural resources is much greater than the European or even world average in that Scotland benefits from 85% of UK hydro-electric energy with a total potential of 1.33 Gigawatts, as well as 25% of European offshore wind power (36.5 GW), 25% of Europe's tidal energy (7.5 GW) and 10% of Europe's wave resources (14 GW). The total energy potential of marine power is thus estimated at 21.5 GW and the total renewable energy potential of Scotland to 60 GW, equivalent to three quarters of electricity generation installations in the UK[2].

In 2011, approximately 35% of Scotland's electricity came from renewables and Scotland contributed almost 40% of the UK's renewable energy output according to Scottish Government estimates[3]. This compares well with other European countries such as France, for example, which only uses 14% of renewables for its electricity consumption. The figure also exceeds the interim target of 31% set by the Scottish Government in 2007. In fact, new targets were introduced in 2009 with the SNP Government's Climate Change Delivery Plan[4] when the 2050 long-term objective was raised from 50% of electricity generated by renewable energy sources to 80%!

In reality, these figures represent a significant economic potential which the SNP has fully integrated to its discourse and which lies at the heart of its economic strategy for an independent Scotland. The think tank Reform Scotland estimates that if this target were to be met, half of renewable Scottish energy could be exported and yield benefits exceeding £2 billion a year[5]. These benefits represent an important potential source of income which the Scottish Government plans to reinvest in an Energy Trust Fund based on the Norwegian model if Scotland were to become

---

[11] http://www.scotland.gov.uk/Topics/Business-Industry/Energy

[12] The Scottish Government, *Scotland Beats 2011 Green Energy Target*, 29 March 2012 (http://www.scotland.gov.uk).

[13] The Scottish Government, *Climate Change Delivery Plan, Meeting Scotland's Statutory Climate Change Targets,* June 2009.

[14] http://reformscotland.com/index.php/publications/details/1260

independent after the 2014 independence referendum. The link made by the SNP between the development of "green" energy sources and a prosperous economy for an independent Scotland is therefore key to the SNP's rhetoric for the current constitutional debate. Not only is the development of renewables in Scotland widely accepted as beneficial to Scotland in environmental terms, but it is presented by the SNP as the key to a prosperous economy for Scotland. According to a 2012 Scottish Renewables study[6], the renewable industry now accounts for 11,000 jobs in Scotland and is bound to create many more with the development of the renewable industry in Scotland. In fact, a 2010 report commissioned by Scottish Renewables and Scottish Enterprise[7] estimated that the offshore wind industry alone could potentially offer 28,000 direct jobs by 2020 in Scotland (supported by a further 20,000 jobs through related industries) if the Government's targets were met, contributing £7.1 billion of investments to the economy. These potential job creations, many of which being much needed rural job creations, would be added to the existing 120,000 jobs that are now being offered by the North Sea energy industry whose expertise could be used for the further development of renewable energy sources and ensure their conversion as North sea oil and gas reserves run out in thirty to forty years.

Such optimistic figures add much needed weight to the SNP's arguments in the current constitutional debate. Indeed, the results of the Scottish Social Attitudes Survey 2011 suggest that 65% of the Scottish electorate is ready to vote favourably to independence if it were to bring them net benefits of around £500 a year, 46% if there was no economic impact whatsoever and only 21% if Scottish independence were to cost them more than £500 a year. The SNP's economic strategy for an independent Scotland is all the more important as the Scottish Social Attitudes Survey 2012 (table 6) found that respondents were equally divided (34% for each option if the results are collapsed) on the Scottish economy's prospects if the country were to become independent. However, if 73% of those who expected that Scotland's economy would get "a lot better" after independence and 45% "a little better" would vote in favour of independence, only 4% of those who expected the Scottish economy to get "a little worse" and 3% of those who expected it to get "a lot worse" would vote favourably to independence (table 7). This suggests

---

[15] Scottish Renewables, *Delivering the Ambition : Employment in Renewable Energy in Scotland*, March 2012

[16] Scottish Renewables, Final Report, *Scottish Offshore Wind : Creating an Industry to Scottish Renewables, IPA Energy + Water Economics,* 4th August 2010.

that the SNP will have to convince a majority of the Scottish electorate that Scotland's economic prospects would benefit from independence for that constitutional option to become a reality after the 2014 planned referendum. Indeed, a majority of the Scottish electorate remains worried (59%) about Scotland's future if it is to become independent as the results of the Scottish Social Attitudes Survey 2012 (table 8) also suggest that 39% of Scots are "quite worried" and 20% "very worried" about Scotland's future, the latter result marking a 5% increase since 2010.

Table 6: As a result of independence, would Scotland's economy become better, worse, or would it make no difference?

	2009	2011	2012
A lot better	6%	10%	6%
A little better	25%	25%	28%
No difference	22%	26%	23%
A little worse	19%	17%	21%
A lot worse	13%	13%	13%
Can't choose	12%	10%	9%

Source: Scottish Social Attitudes 2012, ScotCen

Table 7: Support for independence by expectation of its impact on Scotland's economy, 2012

Expectation of impact on economy	% support independence
A lot better	73%
A little better	45%
No difference	20%
A little worse	4%
A lot worse	3%

Source: Scottish Social Attitudes Survey 2012, ScotCen

Table 8: If Scotland were to become independent, would you feel confident about Scotland's future, worried, or neither confident nor worried?

	2010	2011	2012
Very confident	5%	10%	6%
Quite confident	16%	21%	15%
Neither	23%	22%	20%
Quite worried	39%	32%	39%
Very worried	15%	13%	20%

Source: Scottish Social Attitudes Survey 2010-2012, ScotCen

Moreover, the rosy picture presented by the SNP for an independent Scotland's "green" economy is being increasingly criticized by environmentalists and climate experts as they point out the numerous contradictions inherent to Alex Salmond's plans in terms of energy production. Indeed, the Scottish Government's plans to maximise North Sea oil production over the next forty years with an extraction target of 24 billion boe have been criticized by leading environmentalists for their impact on climate change as it would mean the release of up to ten million tonnes of carbon dioxide into the atmosphere: "*It is clearly indefensible to plan to make Scotland a low-carbon economy but at the same time quite happily export billions of barrels of oil for someone else to burn*"[8]. Scottish green groups and lobbies have thus pointed out that the introduction of low-carbon alternatives and the development of renewables in Scotland should also be accompanied by intensified efforts to reduce demand for fossil fuels, especially as Scotland failed to meet its binding annual target to reduce carbon dioxide emissions in 2010 which rose by 6% to 56 million tonnes. Recent findings of a new study, Delivering Renewable Energy under Devolution[9], published in January 2013, also suggest that many of the Scottish Government's current renewable energy and carbon dioxide emissions targets would have to be revised if Scotland were to become independent as the country would lose much needed UK subsidies for the development of renewable energy. The greater funding costs for Scotland would furthermore increase the cost of Scottish renewable electricity and reduce demand for it on the UK market.

These criticisms tend to undermine the SNP's "green" economic strategy for an independent Scotland and often make it appear as a "green utopia" rather than a credible economic strategy. The use of environmental issues can be considered as beneficial for the SNP in so far as they contribute to its image of a party entirely devoted to defending Scotland's interests, exploit the link between perceptions of the environment and national sentiment, and efficiently serve to stress the inadequacies of the current constitutional setup. However, the economic prospects of an independent Scotland will be key to the outcome of the 2014 independence referendum and the SNP's ambitious programme for a

---

[17] Richard Dixon, director of the environment group WWF Scotland, in Severin Carrell, Scotland's North Sea energy policies irreconcilable with green government, *The Guardian*, 9 October 2012.

[18] The study is led by Dr. Richard Cowell of Cardiff University, Dr. Geraint Ellis of Queens University Belfast, Professor Peter Strachan of Aberdeen University and Dr. David Toke of the University of Birmingham and is funded by the UK Economic and Social Research Council.

"greener Scotland" may prove to be more of a weakness than a strength in the economic debate leading up to it. Indeed, the alternative economic model that the nationalists propose will come under increasing scrutiny. The party's decision to exploit a "green" economic model alongside a more traditional capitalist economic model could prove to be problematic in so far as the inevitable contradictions between the two could be exposed and end up blurring its political message and credibility. Finally, if the SNP's plans and policies for a "greener Scotland" are to be lauded, it is unfortunate that the link made between a "greener Scotland" and an independent Scotland may end up undermining the ambitious environmental policies currently pursued in Scotland today.

PART V:

ECOLOGY AND POLITICS IN IRELAND

IRELAND AFTER THE BOOM AND BUST

FRANK MCDONALD

ENVIRONMENT EDITOR OF THE *IRISH TIMES* AND AUTHOR

The Republic of Ireland is a relatively small country, with a population of less than 4.5 million located at the northwestern edge of Europe; only Iceland is further away. Yet during the late 1990s and throughout much of the first decade of the 21st century, it experienced an unprecedented boom – the so-called "Celtic Tiger"– with economic growth reaching 10 per cent per year or even more at the height of it. The earlier phase was real, driven by manufacturing exports – particularly information technology, pharmaceuticals and, of course, food – but the latter phase was driven purely by construction, fuelled by a range of tax incentives specifically targeted to encourage investment in property. There were lucrative incentives for almost everything – "buy-to-let" apartments, student housing, hotels, run-down seaside resorts and even multi-storey car parks. In most cases, investors could write off 100 per cent of the capital cost of any qualifying project against their tax liability, with the result that many of them paid very little tax, if any.

As a result, construction output soared and property prices reached ludicrously high levels. Annual output of new housing increased from about 30,000 units in 1995 to a staggering 93,000 in 2006 – a rise of 211 per cent; in proportional terms, this was six times the average annual output in Britain. Over the same period, the average price of a house increased by 344 per cent from IR£72,732 in 1994 (before the boom started) to €322,623. Between 1995 and 2007, house prices soared by four times the average construction cost, five times average earnings and seven times faster that the consumer price index. At one stage, in early 2007, the price per square metre of residential property in Dublin was higher than in the Paris; a house on Shrewsbury Road, in Ballbridge – known as the most expensive address in Ireland – was sold for a staggering €58 million. To some of us, it was clear that this was a bubble, inflated by the Government itself and funded (in large measure) by European banks that never seemed to ask themselves what on earth was happening in this small, peripheral EU member state. And, inevitably, it burst, leaving thousands of people in "negative equity", with mortgages worth more than the value of their homes. Several leading property developers accumulated debts of more than €1 billion *each* and the Government – which had been living off the proceeds of stamp duty on real estate transactions – became bankrupt. The most amusing commentary on the whole boom-and-bust saga was artist Fergal McCarthy's installation of giant red and green *Monopoly* pieces on the River Liffey in September 2010.

The country is now littered with "ghost housing estates" – schemes that were never finished after the property market collapsed in 2008, involving a total of 43,000 new homes. And these were disproportionately concentrated in counties Leitrim, Longford, Roscommon and Sligo – all of which were included in an indiscriminate tax incentive programme aimed at encouraging "rural renewal" in the Upper Shannon region. Official figures compiled by the Department of the Environment in 2010 showed that there were more than 2,800 estates in this grim category, some of which represent a danger to the public – particularly children; in February 2012, a two-year old child was drowned in an untended pool of water on an unfinished estate near his home on the edge of Athlone, Co Westmeath. In the wake of this tragedy, the Department allocated €5 million to address the most urgent safety problems. A progress report in July 2012 showed that 211 "ghost estates" – less than 10 per cent of the total – had been resolved. Of the remainder, almost one-third were tied up in legal actions, including enforcement proceedings by local authorities, mainly seeking to recover bonds to ensure their completion.

Ireland is known for its spectacular scenery, particularly along the Atlantic seaboard, thanks to decades of marketing by the State's tourism promotion agencies. But as James Nix and myself showed in our 2005 book, *Chaos at the Crossroads*, the landscape that is the "raw material" of our tourism product was being relentlessly eroded by the proliferation of single houses all over the countryside as well as by inappropriately located schemes, such as the Heritage Hotel and Golf Resort in Killinarden, County Laois, which stands in stark contrast to the agricultural fields around it. There was a time when farmers who built houses for themselves in rural Ireland took care to ensure that these houses nestled into the landscape, often with a shelter-belt of trees to provide protection against the prevailing westerly winds and driving rain. But that was in the 19th century. In the late 20th century, few of those building new homes in rural Ireland were farmers; most, indeed, were simply buying sites to build increasingly large and ostentatious "McMansions" with five or six bedrooms, as many bathrooms and a separate double garage for "his and her" cars.

Throughout the State, and particularly in the West, hedgerows were bulldozed and replaced by suburban-style walls, railings and gates; even the new planting tended to be of the showy suburban variety. And since many of these new houses are located on narrow rural roads several kilometres from the nearest town, cars are essential for getting around – to places of work, shops, churches, schools and so on. Thus, the spread of housing throughout the countryside – which was directly facilitated by the construction of over-scaled motorways – comes with a built-in dependency on cars as the principal means of transport. A map of County Limerick, just south of the Shannon Estuary, showing the number of planning approvals for new houses between January 2000 and August 2003 – a period of just over three-and-a-half years, is literally covered with hundreds of pink dots. Many of the new houses in the West and South-East are holiday homes, also built with scant regard for the landscape or surviving archaeological features. A panoramic view of the Blackwater Estuary, in Youghal, County Cork, has been obscured by a block of holiday apartments that look as if they would be more at home on the Costa del Sol.

Suburban-style housing also spread out from Westport, County Mayo, towards the foothills of Croagh Patrick, which has been a pilgrimage place for centuries. And in Ardmore, County Waterford, the pre-eminence of a characteristic Irish round tower on the site of a monastery founded by Saint Declan has been trumped by a boom-time development of "luxury 4-bedroomed detached houses" on the ridge above it, ironically called "The Heritage". Since Ardmore still has no sewage treatment plant, despite being a traditional seaside resort, unsuspecting visitors are warned by a public notice not to "paddle or bathe in the waters of the stream" running into its harbour, because it is from there that the town's sewage is discharged directly into the sea, on the ebb tide. Trim Castle, in County Meath – acknowledged as the most important Anglo-Norman fortification in Ireland – had its setting gravely compromised by the construction of a hotel right across the street from its medieval battlements; it now trades on having an unrivalled view of the castle.

The construction of new roads has also scarred historic towns such as Sligo, which had the N4 dual-carriageway driven right through its heart, while Clare County Council is trying to get rid of two pivotal mid-19th century buildings in the centre of Ennistymon, merely to make room for a roundabout to cater for tourist traffic on the way to and from the seaside resort of Lahinch. But the biggest crime in this context was the M3 motorway, which was routed through the historic landscape around the Hill of Tara, ancient seat of Ireland's high kings. This dastardly plan was described by poet and Nobel laureate Séamus Heaney as "desecration", while writer Colm Toibín even suggested that the cause of Ireland's current woes could be traced back to the arrogance of destroying the setting of Tara, at a time when we "need places which have a sacred aura and a special beauty more than ever". And it wasn't even necessary; traffic levels on the new motorway are low, mainly due to the economic recession, so taxpayers must compensate the consortium that built the M3 for loss of expected revenue.

The new motorways – in a country that got its first stretch of motorway as recently as 1984 – acted as the sinews of sprawl, permitting commuters to live much further away from their places of work; Dublin's extensive "commuterland" now extends 100 kilometres throughout much of the province of Leinster and even parts of Ulster. It was estimated in 2002 that 70 per cent of the new residents of Gorey, County Wexford, were Dubliners commuting to and from Dublin, nearly 100km to the north. But because there was so much money to be made from the sale of land for housing, the pressures to have it rezoned for residential development proved irresistable to local councillors. This trend towards long-distance commuting was also one of the many downsides of a "decentralisation programme", under which more than 10,000 public servants were to be relocated to 58 different places in 25 counties; it was even to include moving the headquarters of eight Government departments to provincial towns. Mercifully, however, this utterly misconceived scheme was only partly realised.

In Dublin itself, the Docklands area saw feverish levels of construction by a forest of tower cranes during the boom, mostly building new headquarters for firms of lawyers, accountants and other involved in financial services. There were plans for a pair of "skyscrapers", each more than 100 metres high, to form a new entrance to the city from Dublin Bay; one of these was the U2 Tower, which was to have studios for Ireland's most famous rock band on top of it. These plans fell apart because of the property crash, which was dramatically illustrated by the 10-hectare former Irish Glass Bottle Limited site in Ringsend – bought in late 2006 for €412 million by a consortium led by property developer and construction magnate Bernard McNamara and now said to be worth no more than €60 million. The most forlorn symbol of the crash – a skeletal reinforced concrete frame on North Wall Quay, once intended as the headquarters of Anglo Irish Bank, “the bank that broke Ireland” – is also in Docklands. Ironically, this hulk has since been purchased by the Central Bank of Ireland, which had failed miserably to regulate the casino banking culture, as its future headquarters.

During the boom, the dominant thesis among Dublin’s planners was that the city needed “iconic” high-rise buildings, to become like other cities. This was inspired by American pop sociologist Richard Florida’s notion that cities must cater for a “creative class” of highly-mobile young professionals, rather than by any real understanding of Dublin. Even though Google, Facebook and Microsoft – the type of companies that would employ “high bohemians” (in Florida’s terms) – were already here, mainly to avail of Ireland’s 12.5 per cent corporation tax rate, the planners repeatedly approved high-rise schemes that would have profoundly changed the city’s skyline, if they had been built. Most of these plans lie in tatters, including once high-flying developer Seán Dunne’s scheme for a “diamond-cut” 37-storey tower on the Jurys Hotel site in Ballsbridge, a 32-storey residential tower planned for a site just south of Heuston Station and a vast children’s hospital, 16 storeys high and more than 150 metres long, on the Mater Hospital site in Eccles Street. In the case of this behemoth, much to everyone’s surprise, planning permission was refused by An Bord Pleanála, the planning appeals board.

Even though the capital accounts for 40 per cent of the population, there is nobody to speak for Dublin – certainly not in the way that Bertrand Delanoë speaks for Paris, Michael Bloomberg for New York and even Boris Johnson for London; that’s why there is growing support for the idea of a directly-elected mayor with executive power. Civic governance is incredibly weak. Dublin City Council, which should be calling the shots, is a bloated bureaucracy beholden to the Department of the Environment. Apart from its substantial revenue from commercial rates, the council has to seek approval for every cent it spends on a capital investment project, however worthy. The dependence of local authorities for funding from a bankrupt central government, itself beholden to its troika creditors, will continue even after a centralised property tax is finally introduced. (The private sector is also being squeezed by banks not lending because of the EU-ECB-IMF troika’s requirements that they must shrink their balance sheets). Any chance of metropolitan cohesion was squandered when Dublin was arbitrarily carved up between four local authorities -- DCC, Fingal, Dún Laoghaire-Rathdown and South Dublin county councils. A bewildering array of public bodies also exercise power in the city, ranging from the National Transport Authority to the Dublin Port Company.

On a more positive note, at least the boom emboldened many local authorities to commission good architects to design new headquarters for themselves. Prime examples include Fingal County Hall, in Swords, County Dublin, designed by Bucholz McEvoy Architects; Kildare County Council's new Civic Offices in Naas, by Heneghan Peng Architects, and Offaly County Hall in Tullamore, by ABK Architects. Educational institutions also did well, with numerous developments in UCD, Trinity College and Cork Institute of Technology, among many others. But the transformation by Des McMahon of Croke Park from a random collection of stands into an amphitheatre for Gaelic games; that it played host to rugby and soccer internationals was a bonus. The swirling stadium that replaced Lansdowne Road, by HOK Sport Architecture in association with Scott Tallon Walker, is also infinitely preferable to an earlier scheme by former Taoiseach Bertie Ahern for an enormous sports campus (including an 80,000-seat national stadium, instantly dubbed the "Bertie Bowl") at Abbotstown, off Dublin's congested M50 ring road. We've also got two light rail lines and a large number of new bridges, some of them designed to be aesthetically pleasing, as well as new public amenities such as the Liffey Boardwalk. A national conference centre, designed by Irish emigré architect Kevin Roche, finally materialised in Dublin's Docklands after 20 years of haphazard planning, and a new theatre by international "starchitect" Dabiel Libeskind, looks out over the Grand Canal Docks and a public square conjured up by New York landscape architect Martha Schwartz. The new opera House in Wexford, designed by the Office of Public Works and Keith Williams, showed what could be achieved by the public sector at relatively minimal cost compared to, say, the hugely expensive opera house in Olso. In Cork, however, the tallest building in Ireland, the Elysian Tower, stands as a largely empty monument to the madness of the bubble. The Spire in Dublin, a soaring 120m stainless steel needle, can also be seen to commemorate the "Celtic Tiger" boom because it, too, represents nothing. Of all the new developments that encapsulate the Celtic Tiger, Dundrum Town Centre is probably the most potent. Designed by BKD for Joe O'Reilly's Castlethorn Construction, it perfectly expressed Ireland's conversion to Mammon in much the same way that 19th century Gothic cathedrals once symbolised our devotion to God.

The most significant date in recent Irish history was November 29th, 2010, when the Government surrendered our economic sovereignty to the "troika" comprising the EU, European Central Bank and International Monetary Fund in return for a "bailout" of €85 billion – much of it needed to cover casino banking debts. However, despite being in the throes of economic recession, Dublin remains a vibrant city, its central area full of life during the day and at night. Visitors are amazed by the indomitable spirit of its citizens, even in an era of austerity – thus far, at least. Their sense of humour, as exemplified by the IMF's Ajai Chopra being greeted with a cheery "Hello Chopper", sums it up. We should be thankful that the "Celtic Tiger Cub" swagger of the boom years has gone. Chastened by the bust, "Dublin has reverted to being a nice place where there's still a lot going on", as architect Alfred Cochrane observed. We've also managed to maintain much of the city's fabric and prevent it from turning into something like Pittsburgh.

THE POLITICAL DISCOURSE ON NUCLEAR ENERGY IN IRELAND

ALEXIA MARTIN
UNIVERSITY OF TOULON, BABEL RESEARCH LABORATORY

Over the last thirty years, Ireland has emerged as a leader of antinuclearism and has proudly defined itself as "nuclear-free". In 1999, a law even prohibited the use of nuclear power to produce electricity. This policy has attracted wide support across the political divide, which explains the continuity of the non-nuclear policy by successive governments, not only at national level but also within the EU and international organisations.

This objection to nuclear power has mainly focused on the potential risks from Sellafield, the British nuclear site situated on the other side of the Irish Sea, only 120 miles from Dublin. Sellafield was indeed blamed for turning the Irish Sea into "the most radioactive in the world" and for posing unacceptable risks to the Irish population. Hence, the Irish authorities' steadfast opposition and repeated appeal for the closure of the site, which has given rise to tensions between the UK and Ireland.

Nuclear power has provided politicians with an 'easy issue', capable of uniting the Irish around a common nuclear-free ideology. In this respect, the question of nuclear power has participated in the process of identity-building of the young Republic. This calls for an examination of the various aspects of the debate in the Anglo-Irish historical context. Is Ireland's "obsession" with Britain's nuclear industry an extension of her nationalist heritage, in a new form derived from the growing environmental awareness in industrialised societies? Some answers will be provided here using an analysis of the political discourse and response to the risks posed by Sellafield, which have sometimes given rise to media posturing and political exploitation.

The pro-nuclear discourse of the 1970s

In the middle of the 1960s, the Irish had their own civilian nuclear plans following the example set by their British neighbours. In 1968, the Fianna Fáil Government chose the site of Carnsore Point, a headland in the very South East corner of County Wexford, for the construction of its first nuclear power station. The intention was to open up to the world stage – after a decade of protectionism – and to join the industrial modernity of the leading European countries. According to Alan Stanbrook in *The Economist* of April 1977, Ireland was then at a crossroads in its history:

> Will it continue to pay lip service to the narrow, provincial aspirations more appropriate to 50 years ago than to the last quarter of the twentieth century? Or will it earn its place as a fully mature member of the EEC?"[1]

The answer was obvious for personalities from the world of science, as well as from political and economic circles: nuclear power was an economic and strategic imperative, an important piece of the puzzle for pursuing the great change. An example of this stance was given by opposition MP Barry Desmond reminding the Dáil of the national priority:

> All political parties are concerned about economic growth and industrial expansion and these twin objectives are the cornerstones of the policies of all the political parties. Inevitably, these policies mean that we require massive growth in energy consumption.[2]

This consensus was the single-most important reason for the lack of substantial political debate on the nuclear project until the middle of the 1970s. The first objections were raised locally, then nationally, following the arguments put forward by antinuclear and peace movements in other European countries. More doubts were raised after the accident at the Three-Mile Island power station in the United States, in March 1979.

The Irish antinuclear movement reached its climax in August 1978 in a joyful summer festival on the site of the government's nuclear project, Carnsore Point. For three days, some 25,000 people gathered to listen to antinuclear speeches and to attend theatre and traditional music performances. While the festival was continued every summer until 1981,

---

[1] Alan Stanbrook, "Survey: Ireland", *The Economist*, 9 April 1977.

[2] Dail Eireann, "Parliamentary Debates", Official Report, vol.303, n°10, Dublin : Stationery Office, 21/02/1978

the antinuclear movement developed all around the island, with about fifty local protest groups in Eire and Northern Ireland.

The government did not easily give in to popular pressure. The Secretary of Industry, Commerce and Energy, Desmond O'Malley, fiercely stood up for the nuclear project. Whereas all the key players of the debate asked for a public inquiry on the issue, he still argued that "a public inquiry in relation to a technology which is already being successfully applied in many countries would hardly be a helpful exercise".[3]

Following the 1981 General Elections, the new coalition government (Fine Gael/Labour) announced that the project would not be pursued as long as the "environmental issue" remained. Thereafter, the nuclear option permanently disappeared from the political agenda and antinuclearism became the norm in politics and public opinion.

It may therefore be asked how in a brief six years, Irish politicians went from a firm pro-nuclear stance to a complete rejection of the atom, not only in Ireland but also on the British coast, with the strong opposition to Sellafield. The Chernobyl catastrophe in April 1986 certainly raised awareness among public authorities of the way radioactivity spreads across national boundaries. Nevertheless, we can assume that popular pressure largely contributed to this U-turn and that public opinion dictated the successive governments' nuclear policies. In fact, this is not peculiar to Ireland. Sociologists have observed that since the 1970s, environmental issues have been raised by NGOs first and not by scientists or public authorities. Similarly, after the debate on Carnsore Point, it was mainly protests from environmental groups that led the government to condemn the British management of Sellafield.

The anti-Sellafield consensus

The mobilization against Sellafield emerged in the early 1980s in County Louth, on the East coast of Ireland, just opposite Sellafield. It started with an article published in the prestigious *British Medical Journal* and showing a link between the 1957 fire at Windscale (the original name of Sellafield) and an unusual cluster of babies born with the Down's syndrome.[4] Local environmental groups and politicians were also denouncing the impact of radioactive emissions from Sellafield, allegedly

[3] Quoted by Matthew Hussey & Carole Craig, *A Nuclear Ireland?,* Dublin : Coop Books, 1978, p.38.

[4] Sheehan, P.M.E and Hillary, I.B, "An unusual cluster of babies with Down's syndrome born to former pupils of an Irish boarding school", *British Medical Journal,* 287 (12/11/1983), pp. 1428-29.

responsible for a high rate of some cancers and congenital diseases in the region. That is the reason why the local Sinn Fein MP Arthur Morgan has deplored "a near cancer epidemic" affecting the East coast and particularly County Louth.[5]

In 1994, this led four inhabitants of Dundalk to launch legal proceedings against the new Sellafield reprocessing plant, THORP.[6] Indeed, they decided to sue not only the public operator in charge of Sellafield (BNFL), but also the Irish State, "for failing to protect the plaintiffs and Irish citizens from the hazards of THORP by not taking appropriate steps to prevent its operation".[7]

The case was widely reported in the media, yet coverage of the issue only peaked a few years later, after 9/11, when an attack on Sellafield was feared following explicit threats from Al-Qaida to strategic British sites. At that time, Greenpeace and a number of key Irish figures carried out several "Shut Sellafield!" campaigns.

The widespread political condemnation of Sellafield is obvious when skimming through the numerous parliamentary debates on the issue. As early as 1986, former Prime Minister Charles Haughey blamed the British authorities for trying to cover up cases of child leukaemia at Seascale, a village directly adjacent to the nuclear site. He then called for a closure of the site and the charging of the people "who have clearly lied to us about this case for the last four or five years".[8]

The issue went beyond political divisions and has sometimes been a factor of national unity. This was illustrated by the coalition government (the "Rainbow Coalition", Dec.1994-Jun.1997) formed without elections after the scandals affecting the previous government. Taking Sellafield as a national cause, they immediately announced a diplomatic plan against the new reprocessing plant, THORP. The idea was clearly to unite the

---

[5] *The Irish Independent,* 27 August 2003.

[6] THORP for Thermal Oxide Reprocessing Plant. Spent nuclear fuel from reactors is reprocessed to separate the 96% uranium and the 1% plutonium, which can be reused in mixed oxide fuel, from the 3% radioactive wastes, which are treated and stored at the plant. The uranium is then made available for customers to be manufactured into new fuel. The plant was given the go-ahead in 1978, although it was not completed until the 1993 at a cost of £1.8 billion.

[7] "Ireland and the Attorney General are also being sued for failing to protect the plaintiffs and Irish citizens from the hazards of THORP by not taking appropriate steps to prevent its operation", Stop THORP Alliance Dundalk (STAD) website, http://stad.ie (12/10/ 2005).

[8] *The Guardian*, 30 July 1984.

Irish around a common plan, with nationalist leanings capable of bringing the country together.

If most election manifestos already disapproved of Sellafield in the 1990s, it became a key issue in the aftermath of 9/11. "It would be a political suicide not to be against Sellafield", Trevor Sargent, the leader of the Green party, admitted in May 2005.[9]

Although all parties agreed on content, there were many criticisms on the lack of governmental action despite rhetorical commitment. This accusation reached a climax during the legal action by the four Dundalk residents. In 1997, when the media revealed that the government was reluctant to fund the legal proceedings partly – as promised during the election campaign – the opposition parties expressed outrage, talking about a "U-turn" and "hypocrisy". The authorities eventually kept their word, yet the controversy continued, and it was argued that the government was "shamed into action".[10]

In reality, the government's determination has increased constantly. In the years after 9/11, some British commentators saw in this escalation a sign that the stance of the Irish was radicalising. Invited at the annual Congress of Fianna Fáil in 2002, a top manager of Sellafield thus warned against "a dangerous consensus" among Irish political leaders, which could only be detrimental to dialogue:

> It allows for no reasonable critique as to what is desirable, or practical, fair or balanced. […] It could only mean that the political parties and their leaders are locked into a competition as to whose anti-nuclear credentials are the best.[11]

We must admit that many politicians vied with anti-Sellafield statements in the media. Each decade had its "stars" of antinuclearism, politicians or performing artists: among others, Charles Haughey (Fianna Fáil) in the 1980s, Emmet Stagg (Labour party) and April Doyle (Fine Gael) in the 1990s, European Green MPs Patricia McKenna and Nuala Ahern, Senator Shane Ross in the 2000s, not to mention U2 singer Bono and his wife.[12]

---

[9] Interview, Alexia Martin, May 2005.

[10] "Shamed into action", *The Evening Herald*, 16 October 1997.

[11] John Clarke (in charge of quality, security and environmental matters at Sellafield), "What Next for Sellafield", Speech to the Fianna Fáil Dublin Forum, 10/12/ 2002.

[12] "I was the loudest voice in the Senate against the dangers to Irish people from Sellafield", Senator Shane Ross, "Election Address", June 2002.

It may be tempting to say that politicians used the media to stage their own publicity, thanks to a consensual issue that draws upon some "obscure" nationalist feelings. Indeed, most anti-Sellafield statements seem to address the voters more than the British authorities. Yet, there has been little debate on a possible political exploitation of the issue, aside from a few voices on the margin. One of them is Damian Byrne, an editor of the *Irish Examiner* who was highly critical of the various anti-Sellafield campaigns. In April 2002, his tone clearly went against the flow, at a time when the anti-Sellafield postcard campaign by Ali Hewson (Bono's wife) was raging throughout Ireland:[13]

> The Sellafield issue also holds obvious attractions for our politicians. Ever since the Nice Treaty debacle, Ireland's political elite have been behaving like a bunch of angst-ridden teenagers, wondering how they have managed to become so out of touch with the public. The forum on Europe road show was devised to help get them back in touch with "the people". But then came the abortion referendum to demonstrate once again, what little hold they have over us. In this context, Sellafield offers politicians that increasingly rare commodity – an issue on which they are sure of being in touch with popular opinion and on which they can engage in as much populist breast-beating as they like without the risk of offending any constituency.[14]

Whether he is cynical or clear-sighted, Byrne was not the first to mention the importance of demagogy and vote-catching practices in Irish political life.[15] However, it would be unfair to deny the sincerity of some elected representatives and to sweep aside the genuine fears as to the risks from Sellafield. Besides, some politicians have not hesitated to go personally to Sellafield or to engage in direct action with campaigners.[16]

---

[13] It consisted in sending a ready-made protest postcard to Sellafield CEO, Tony Blair or Prince Charles. One million were sent via the national post, which did not charge for stamps.

[14] Damian Byrne, "Consider the evidence before you send off those Sellafield postcards", *The Irish Examiner,* 18 April 2002.

[15] Padraid Kirby, *Has Ireland a Future*?, Cork & Dublin: The Mercier Press, 1988, pp.86-91; Michel Peillon "Ireland: A political system apart", *Etudes Irlandaises*, Dec. 1990, p.133.

[16] For example, European MP Nuala Ahern joined activists to "welcome" Mox freighter in Barrow-in-Furness in September 2002, as a protest against the transport of radioactive substances on the Irish Sea.

Antinuclearism or nationalism?

The Irish accused Britain of only considering its own interests, as illustrated by this angry statement by Senator Mary Kelly in 1997:

> It would seem that what is driving Nirex, THORP and BNFL is money and if that is their only concern then they will sell plutonium to the highest bidder without worrying about the consequence.[17]

The poor communication on this issue can be explained by the different priorities underlying the controversy, which have been economic in the UK and environmental in Ireland. This is reflected by the two different government departments dealing with Sellafield in each country: Industry or Energy in the UK and Environment in Ireland. It should also be noted that most Irish politicians showed only a partial knowledge of the operations on the site, which did not facilitate discussions.

Be that as it may, meetings on the issue have not always been easy, such as the ministerial meeting of May 2003 between Martin Cullen and Brian Wilson.[18] The latter left quite upset, condemning a "dishonest" debate and the "long procession of unfounded scare stories" in the media concerning the controversial MOX plant.[19] These difficulties can easily lead one to see the conflict as somehow connected with the age-old Irish fight against British imperialism. In that case, the Irish approach would stem from a symbolic transfer of past rivalries, and not from purely environmental concerns. Nuclear power would then represent a new form of colonization by vengeful neighbours.

Of course, the situation is more complex than it seems. First, it is important to remember that there is a real historical and ideological

[17] Seanad Eireann, "Sellafield Nuclear Processing Plant: Motion", volume 149, 1716-1717, 06/02/1997.

[18] Respectively, the Irish Minister for the Environment, Heritage and Local Government and the UK Energy Minister.

[19] *The Irish Times*, 29 May 2003. From 2005, discussions between Dick Roche and Alan Johnson were far more courteous and fruitful.

The MOX plant was set up to create mixed-oxide fuel (MOX) for use in nuclear power plants, with its chief customer being the Japanese nuclear industry, including the Fukushima complex. The plant was built in 1996 and became operational in 2001. It was closed down in August 2011, following the Fukushima accident. The setback casts doubt on the viability of reprocessing, with its growing mountain of reprocessed waste and provides further evidence of the changed international nuclear picture.

consistency in the Irish antinuclear discourse. Antinuclearism could even be related to a "tradition". Indeed, it is in line with two founding principles of the Irish Republic:

- The policy of neutrality, because of the obvious links between military and civilian nuclear programmes
- The role of Ireland in the Non-Proliferation Treaty of 1 July 1968, insofar as it was largely due to the determination of Irish Foreign Affairs Minister Frank Aiken that the issue of nuclear proliferation was eventually put on the UN agenda, which led to this major agreement for world-peace.

Given this role for worldwide denuclearization and the importance of the policy of neutrality in Ireland, it was quite consistent that Ireland would oppose any potential risks from nuclear sites, on its territory or from nearby Sellafield. This continuity tends to show that antinuclearism is one of the key elements contributing to the Irish distinctiveness and sense of identity.

Nevertheless, the fact that Sellafield is run by the former British colonial power could hardly be neutral. At the height of the mobilization against Sellafield (mid-90s - early 2000), antagonisms in Northern Ireland were still alive – despite significant progress after the 1998 Good Friday Agreement. For that matter, it was surprising to observe the passionate and even vociferous political and media response to Sellafield in Ireland. One would say that the nuclear issue is usually a sensitive one but there is more to it. The arrogance sometimes shown by the British authorities in response to Irish fears might have stirred up old resentments on the Irish side. When I asked Frank McDonald, the environment editor of the *Irish Times*, if there was an anti-British feeling in media coverage in Ireland, he said:

> Possibly, possibly not. I think we have overcome a lot of traditional 'anti-British' sentiment in Ireland over the past few decades, and that's a good thing. But you have to ask yourself why they chose a site for Windscale that was the closest to us, other than Wales. If the whole operation is so safe, why couldn't it be in, say, Hammersmith?[20]

In Ireland, the Sellafield controversy was frequently linked to *The Troubles* in Northern Ireland. For instance, shortly after the Joint Declaration of December 1993, a parliamentary debate raised the possibility

[20] Interview, Alexia Martin, May 2005.

of putting Sellafield on the agenda of the peace talks.[21] It turned out that the issuing of this Joint Declaration coincided with the commissioning of the THORP plant, the much decried reprocessing plant. Some saw the concomitance of both decisions as evidence of backroom dealings between British and Irish governments.[22] Was Sellafield a bargaining chip in talks on Northern Ireland? It is obviously difficult to say... However, if there had been such dealings, the Irish government would have suddenly become less critical about Sellafield. On the contrary, the Fianna Fáil government pursued a much more offensive approach. This new tone is best exemplified by Taoiseach Bertie Ahern's strongly worded declaration, as the Sellafield MOX plant was about to be put in operation, a few weeks after 9/11:

> It is not acceptable that the Irish Sea is used as a kitchen sink by the nuclear industry. The position of Fianna Fáil is clear and uncompromising. We demand that Sellafield is shut… The issue will not be allowed to rest. The campaign to close Sellafield has only begun. It is a campaign that will be waged relentlessly until it is won.[23]

One month later, the governing party issued a similar statement, though using a different and original strategy. In a full page purchased in the British daily newspaper *The Times*, Bertie Ahern and his fellow party members addressed British citizens directly:

> Shut Sellafield!
>
> Sellafield poses an unacceptable and unnecessary risk to our environment. It has contaminated the seas and environment of Ireland and Britain with radioactive discharges for more than forty years. It continues to accept the nuclear waste of the world for reprocessing. This process creates nuclear waste so toxic, that it must be kept away from human contact for hundreds of thousands of years. From the Windscale fire in 1957 to the falsification of safety records in 2000, and the continued pollution of our air and seas, Sellafield has posed an unacceptable threat to the environment and security

[21] John Major, then British Prime Minister, and Albert Reynolds, then Taoiseach, issued a joint declaration from 10 Downing Street, London, which became known as the *Downing Street Declaration* argued for self-determination on the basis of consensus for all the people of Ireland.

[22] According to John Bowler from Greenpeace Dublin, this was no coincidence. As other anti-Sellafield activists, he feared that the Irish government would be less vocal in opposing the THORP plant because of a secret agreement. See also: "THORP approval linked to peace declaration?", *The Cork Examiner*, 16 December 1993.

[23] Fianna Fáil Ard Fheis (annual conference), 12/10/2001.

> of these islands. In the light of this threat, we call on the British government to reverse the decision, taken in October, to proceed with the commissioning of the Mixed Oxide Fuel (MOX) plant at Sellafield. [...] Furthermore, in the aftermath of the September 11 assault on Manhattan and the Pentagon, we also believe that Sellafield poses a grave security risk to both our countries. A similar assault could render much of our Islands uninhabitable. [...]
>
> We, members of the Fianna Fáil parliamentary Party, the party of Government and the largest political Party in Ireland, call on the British Government, to reverse their decision on MOX and to shut the Sellafield complex.[24]

Having obtained no response to this solemn appeal – and spurred by popular mobilization – Ireland initiated legal proceedings against Britain in two international Courts of Justice (The Hague and United Nations).[25] The objective was to halt the production of MOX through objecting to the non-disclosure of crucial information about MOX transport, terrorist threats and pollution of the marine environment. This led to an Anglo-Irish agreement stipulating that the British should share some information with the Irish and let them visit the site regularly.

Other elements suggest a link between anti-imperialism and antinuclearism, such as the fact that the republican/nationalist vote is traditionally among the highest in counties Louth and Wexford, which are also known for their antinuclear activism[26]. Moreover, the Scottish nationalists have regularly endorsed the Irish claims over Sellafield, as in this statement by MSP Bruce Crawford (SNP) at the 2002 British Irish Council (BIC):

> The meeting of the BIC today is an excellent opportunity for Scotland to stand shoulder to shoulder with Ireland in a bid to reverse the UK government's decision to have Mixed Oxide Fuel (MOX) production at Sellafield in England. Today I have lodged a parliamentary motion supporting the Irish government in their legal case. Clearly, the Irish treat the threat of nuclear pollution coming from a normally friendly neighbour very seriously. [...] It is important for a small nation like Scotland to stand by Ireland on such a vital environment issue.[27]

---

[24] *The Times*, 24 November 2001.

[25] International Tribunal of the Law of the Sea (ITLOS), United Nations, October 2001; Permanent Court of Arbitration (PCA), The Hague, October 2002.

[26] County Louth is the closest from Sellafield and the people of county Wexford were the first to oppose the Carnsore Point nuclear plan.

[27] "Parliament should support Irish in legal case", site internet du SNP, 22/11/2001 (http://www.snp.org/pressreleases/2002/news.695/).

This stance tends to support the view of an anti-imperialist feeling underlying the dispute.

Having observed a historical consistency in the Irish political discourse on nuclear energy, it can be assumed that antinuclearism helped Ireland to assert its distinctiveness, especially *vis-à-vis* the former colonial power. In fact, from a historical perspective, Irish identity has developed drawing on a logic of opposition to intruders, of rebellion. Similarly, nuclear power has come to symbolize 'The Other' for Ireland, an entity which tries to impose its model, its rules and values. That is the reason why opposition to Sellafield may well be viewed as a means for the Irish to reclaim their territory, in the face of the legacy of colonialism, the opening to foreign capital and European homogenization.

However, what seems to be a firm antinuclear discourse contains a number of flaws, which somewhat undermine this stance:

- A poor environmental record, with repeated infringement of European directives;
- Insufficient energy supply, which fuels a growing Irish pro-nuclear lobby;[28]
- An ambiguous European policy, in which Ireland comes out as antinuclear (one of the few before Fukushima) while co-funding the European Atomic Energy Community (Euratom), which is a moral dilemma according to the Green Party.

Since 2003, there has been a deepening disinterest for Sellafield in the Irish media and the political discourse. This can be explained by the announcement of the closure of THORP for 2010 and by a better collaboration between the British and Irish authorities on nuclear safety. Since the recent economic downturn (the worst crisis the country has known), political matters have mainly focused on the critical economic situation of the country.

The Fukushima disaster has renewed fears of nuclear risks from Sellafield, though only weakly. Ireland has officially asked that the site be part of the stress tests of the European nuclear sites.[29] Despite the assurance of the British Environment Minister Phil Hogan, Britain has

[28] The most vocal pressure group on nuclear power is Better Environment with Nuclear Energy (BENE), http://www.bene.ie

[29] Following the nuclear accident in Fukushima, the EU agreed on a series of voluntary tests on its nuclear power plants, based on a set of common criteria.

finally withdrawn Sellafield from the tests.[30] But as it is one of the sites chosen for the relaunching of the British civilian nuclear programme, the dispute might well be relaunched soon enough…

[30] The argument given by the British authorities was that Sellafield did not produce nuclear power, which overlooks the storage of highly radioactive substances from reprocessing.

THE COLOR GREEN IN IRELAND: ECOLOGICAL MYTHOLOGY AND THE RECYCLING OF IDENTITY

FRÉDÉRIC ARMAO
UNIVERSITY OF TOULON, BABEL RESEARCH LABORATORY

> "He had that curious love of green, which in individuals is always the sign of a subtle artistic temperament, and in nations is said to denote a laxity, if not a decadence of morals."
> Oscar Wilde

The color green is deeply connected with various, clearly distinct concepts and ideas in our modern world and its symbolic role seems multifaceted to say the least: if it is connected for obvious reasons to Nature –and has been so for centuries[1]– the 19th and 20th centuries associated the color with numerous aspects of our social life, be it the concept of ecology and the environment, money and capitalist values – as the color is still to this day used on dollar bills – or more political and religious matters, from the ancient association to Islam, which is today perhaps more lively than ever, to the modern creations of a number of national flags, including the *bratach na hÉireann*, i.e. the national flag of Ireland.

The connection between the color green and the very idea of Ireland itself seems prominent today. The Emerald Isle may take its nickname from its lush verdant vegetation which has inspired a large number of poets throughout the centuries; according to the Christian tradition, this naming might have actually been inspired by Saint Patrick himself –although this claim does not stand up to historical analysis. Many questions remain unanswered and this study aims at analyzing the role

---

[1] See for instance PLINY THE ELDER, *Naturalis Historia*, 37, 62. Quoted by TRINQUIER Jean, "Les Vertus Magiques et Hygiéniques du Vert dans l'Antiquité", in *Couleurs et Visions dans l'Antiquité Classique,* Rouen, Publications de l'Université de Rouen, 2002, pp. 97-128.

played by the color in Irish history and more specifically its impact on contemporary Irish identity: the use of the color in itself –more than the use of the mere term "green" in connection with Ireland– will therefore be studied. I will first of all wonder whether the association between the color and Ireland is justified and relevant while reviewing the different theories that may or may not explain its apparent recurrence in connection with Irish history; in other words, the question will be: to what extent is Ireland inherently "green"? The use of the color in the context of modern Irish society will then be analyzed in order to understand how exactly the color contributes to the building of contemporary Irish identity (or identities):[2] our focus will include a study of Irish folklore –more specifically that of Saint Patrick's Day. This research will finally concentrate on the political use made out of this color in Ireland which, as can be easily imagined, goes far beyond the concept of ecology alone. This study will therefore try and provide some answers and perspectives on the following questions: is green actually the color of Ireland? To what extent can it be considered a part of the Irish identity? How and to what end is it used today in a political and ecological context?

From a strictly political point of view, the answer to the question "is green the color of Ireland?" is rather simple: it is not, as no legal document mentions the formal use of an official color for the Republic of Ireland. As a matter of fact, if there were to be an official color, it would most likely be the color blue; the *Standard of the President of Ireland*, that is to say the official flag which has been used by the Irish president since 1945,[3] is similar to the coats of arm of Ireland: it represents a golden *Clàrsach* (the Gaelic harp) on a blue background. The name of this blue color, which is also to be found in the Irish heraldry, remains to this day "Saint Patrick's Blue" and the color seems to have been associated with the island since at least the tenth century:

> The State harp is invariably shown on a deep blue background. Here it is worth noting that *Gormfhlaith* appears in the early Irish texts as the name of several queens closely connected with dynastic politics (including the Kingship of Tara) in the 10th and 11th centuries. *Gormfhlaith* is a compound of gorm (blue) and flaith sovereign. In early Irish mythology the

---

[2] Regarding this debate, see for instance GUILLAUME André, *L'Irlande, une ou deux nations ?*, Paris, P.U.F., 1987.

[3] 9 November 1945, National Library of Ireland. Genealogical Office: Registrationf of the Arms of Ireland. See http://www.nli.ie/GetAttachment .aspx?id=72fd744a-aced-43b4-a63b-b64be7b272ec.

sovereignty of Ireland (*Flaitheas Éireann*) was represented by a woman often dressed in a blue robe.[4]

Henry VIII officialized the Irish coats of arms –the golden harp on a blue background ("d'azure à la harpe d'or")[5] – at the creation of the Kingdom of Ireland in 1542[6] and the symbol is still used on the modern version of the Royal coats of Arms of the United Kingdom in order to represent Northern Ireland.

This fairly clear historical background seems at odds, however, to our contemporary "collective imagery" or rather to the modern public Western imagination which seems to connect the country of Ireland to the color green. Yet stating that "most people" associate the color green to Ireland in the 20th and 21st century is one thing: proving it is another matter. There is to our knowledge no thorough scientific research demonstrating that the color green is associated to Ireland by a majority of people. We have just shown with the example of Saint Patrick's Blue that Ireland is not inherently "green": why would this color be universally acknowledged as a modern symbol of the island of Ireland? Hopefully, the modern researcher has at his disposal powerful tools and sometimes the simplest ideas might be the most effective in trying to prove a point. Thus, simply typing the word "Irish" into a popular search engine[7] – a practice often referred to as the "Google test"– showed a rather effective way of establishing the current association of the color to the demonym. The 850 first pictures found by the search engine undoubtedly favor the color green over any other (blue or orange[8] for instance) –a fact confirmed by a color histogram analysis of those 850 images.[9] Although by no means a strictly scientific process,[10] this approach enables us to acknowledge the fact that,

---

[4] *Ibid.*, p.4.

[5] This description is to be found in a French roll of arms dating back to the 13th century. O'LAUGHLIN Michael C., *The Irish Book of Arms*, Irish Genealogical Association, 2001, p. 15.

[6] See *The Parliamentary Gazetteer of Ireland 1847-1848*, Vol 1, Dublin, 1850 quoted by John Kennedy. See http://www.theheraldrysociety.com.

[7] http://images.google.com.

[8] Those two colors were analyzed and compared to the use of the color green for obvious reasons as they could stand as a symbol of Ireland: blue for the blue of Saint Patrick and orange for the other color of the national flag of Ireland.

[9] The detail of this analysis and the meta-pictures of the images put forward by Google can be found here: http://bit.ly/OR74Im.

[10] The results produced by a "Google test" are not absolute in that they cannot but reflect indirectly what web-surfers and web-masters (a fraction of the human population in itself) searched during a given period of time and by no means what

today, a substantial fraction of the population (people using the Internet, creating websites and implementing pictures) clearly associate the color green to Ireland and that this association is sufficiently powerful to produce images which look predominantly "green".[11]

The color is therefore – at least partially – associated to Ireland. Before this research moves to political and ecological matters – i.e. to what extent this "Irish color" is used today in those political and ecological context– one may simply wonder the reason why the color green is so closely connected to Ireland. The answer seems, as always, obvious: the island is, factually speaking, "green". In other words, the green landscapes of Ireland might have triggered this association with the color and, among other examples, the popular nickname of "the Emerald Isle" is quite relevant. The creation of the term "Emerald Isle" is generally attributed to William Drennan in his late 18th-century poem "When Erin First Rose";[12] despite the highly political nature of the poem it is true that the green aspect of the island necessarily played a role in this association and most likely coined the term "green Erin".[13] The country is today one of the least wooded in Europe as it was massively deforested, especially at the time of the English settlement and the plantations of the 16th and 17th centuries. However, the mild humid climate of the island still favors lush vegetation of herbs and grass. If Ireland is in fact green, the Irish "green" would then be that of Nature –a theme which will later be taken up by the Irish ecologists. The association between the color green and the beautiful landscapes of Ireland most likely appeared along with –or was at least

a majority of humans think, feel and imagine regarding a specific word. Therefore those results are necessarily biased and partial and they obviously need to be interpreted and contextualized.

[11] The definition of a "color" as an optical phenomenon (what is green and what is not) is in itself problematic; we assume that this research does not need to go into detail relative to optical sciences –such as references to the spectrum and the proper wavelengths corresponding to the color green.

[12] "Let no feeling of vengeance presume to defile / The cause of, or men of, the Emerald Isle." *The cabinet of Irish literature,* READ Charles Anderson (ed.), London, Blackie & Son, 1884, Vol.2.

[13] "When Erin first rose from the dark swelling flood, /God bless'd the green island and saw it was good". But once again, the poem is before anything political: "O! sons of green Erin, lament o'er the time / When religion was war, and our country a crime" or "The cause it is good, and the men they are true / And the Green shall outlive both the Orange and Blue." See *supra* for the political use of the color.

strengthened by– the Romantic Movement.[14] However, from a purely societal and political point of view, the first connection between the color and the country appeared undoubtedly before.

Between 1641 and 1653, the Green Harp Flag was one of the symbols of the Irish Catholic rebellion against English Protestant settlers during the Irish Confederate Wars; there is in particular a mention of its use by Eoghan Ruadh Ó Néill, one of the leaders of the rebellion[15] and the Green Ensign was apparently used by various Irish merchants from the 17th century onward.[16] During the Irish Rebellion of 1798 a version of the Green Harp flag was once again used by the rebels: the idea was to change the golden harp on a blue field, a flag which had officialized by Henry VIII, and turn the blue background into green in order to represent the change of sovereignty – a shift from the English supremacy to independence. In other words, the color green was used as a symbol of independence and rebellion as early as the 17th century in Ireland and more specifically a representation of the Catholics against the Protestants.[17] The question however remains: why should green represent Ireland and the Irish rebellion in general and Catholicism in particular? Two possible explanations can be put forward and they are both connected to some extent to the notion of Nature. The first answer comes directly from the Irish rebels in 1798. The oath[18] of the United Irishmen –obviously united against the English– runs as follows:

> What have you got in your hand? - A Green bough.
> Where did it first grow? - In America.
> Where did it bud? - In France.
> Where are we going to plant it? -In the crown of Great Britain.[19]

---

[14] It is hard to track the first instance of the use of the term "green" associated with Nature in Ireland; this study focuses of the use of the color in itself, not on the use of the term.

[15] HAYES-MCCAOY G.A., *A History of Irish Flags from the earliest times*, Boston, G.K.Hall&Co, 1979.

[16] http://www.crwflags.com/fotw/flags/ie-colon.html See also a document provided by the official website of the Prime Minister of Ireland: http://www.taoiseach.gov.ie/eng/Youth_Zone/About_the_Constitution,_Flag,_Anthem_Harp/The_National_Flag-PDF.pdf.

[17] Eoghan Ruadh Ó Néill was for instance one of the strongest proponents of Catholicism in 17th-century Ireland.

[18] Also called "catechism" of the Irish Ulstermen. See *supra*.

[19] PAKENHAM Thomas, *The Year of Liberty. The Great Irish Rebellion of 1798*, London, Hodder & Stoughton, 1969.

This oath does not only place the Irish rebellion in the context of some the most notable political events of the 18th century, that is the American independence and the French revolution:[20] it is also the first instance of an official political metaphor based on "natural"[21] (in the sense of "relative to Nature") grounds. This metaphor of a growing and flourishing green bough of liberty could therefore be one of the origins of the popular use and association of the color in Ireland. It should be noted that during the rebellion of 1798, the mere fact of wearing green ribbons, garments or sometimes a shamrock –the famous three-leaf clover – was not tolerated by the English authorities as it implied that the wearer was a supporter of independence.[22] A 19th-century version of the popular Irish ballad *The Wearing of the Green* is rather telling:

> They're hanging men and women there / For wearing of the green.
> Then since the color we must wear /Is England's cruel red
> Sure Ireland's sons will n'er forget / The blood that they have shed.
> You may take the shamrock from your hat /And cast it on the sod, But
> 'twill take root and flourish still / Tho' underfoot 'tis trod.[23]

Consequently, the color green and the shamrock both represented not only the rebellion but also one key figure of Irish history: Saint Patrick. And indeed, the other possible explanation for the political and religious use of the color might be even older than the 17th or 18th century – and paradoxically may have even more impact on modern politics. According to tradition, Saint Patrick is the Apostle of Ireland and is today one of the most important figures of Irish Catholicism. The Saint is supposed to have Christianized Ireland in the 5th century[24] and is believed to have used a shamrock, a three-leaf clover, in order to explain the concept of Trinity to the heathen Irish. Even if this legend is a modern reconstruction (as there

---

[20] Some authors even believe the green bough to be a sprig of acacia, inspired by Freemasonry. CAULFIELD James, *The Great Irish Rebellion, Secret and Silent Men of 1798*, Victoria, Trafford, 2004, p. 114.

[21] The terms "environment" and "ecology" are obviously difficult to apply in an 18th-century context.

[22] ZIMMERMANN, Georges Denis, *Songs of Irish rebellion: Irish political street ballads and rebel songs, 1780–1900*, Dublin, Four Courts Press, p. 167.

[23] "The Wearing of The Green" by Dion Boucicault.

[24] The history of the Christianization of Ireland is however much more complex. See MAIGNANT Catherine, *Histoire et Civilisation de l'Irlande*, Paris, Nathan, 1996, pp. 7-13.

is no mention of such events at or around Saint Patrick's time),[25] the Saint was actually associated to the concept of Nature through the prism of the now worldwide famous shamrock at least as early as the 1680s.[26] This is accentuated by the fact that Saint Patrick is traditionally celebrated at the beginning of spring, on the 17th of March, and that most of his legendary feats are supposed to have taken place at or around Easter, i.e. at a time where Nature renews itself and vegetation regenerates.[27] Saint Patrick is therefore directly associated to Nature: on the one hand to the three-leaf clover and on the other to a flourishing Nature that is brought back to life at the beginning of spring. The fact that Saint Patrick is often associated with the color green –beside the existence of an "official" Saint Patrick's blue– should therefore not be surprising at all: the Saint can often been seen holding a green Shamrock, wearing a green chasuble and/or surrounded by greenery on the numerous representations that are to be found, for example, on stained-glass windows of churches and cathedrals.[28] By extension, the color green became over the centuries the symbol of Irish Catholicism and therefore of the struggle against the English settlers and later the British army. This gave birth to the Irish flag as we know it today, as it was created in 1848,[29] adopted in 1916 then officialized in 1937;[30] the official website of the *Taoiseach*, the Irish Prime Minister, acknowledges the fact that:

---

[25] MACKILLOP James, *Dictionary of Celtic Mythology,* Oxford, Oxford University Press, 1998, pp. 363-4.

[26] See CRONIN Mike and ADAIR Daryl. *The Wearing of the Green, a History of Saint Patrick's Day*, London, Routledge, 2006, p.25. Consider also the following quote: "This Plant is worn by the People upon St. Patrick's Day. It being a current Tradition, that by this Three Leafed Grass, he emblematically set forth to them the Mystery of the Holy Trinity." Threlkeld Caleb, *Synopsis Stirpium Hibernicarum*, 1726 at "Trifolium". Quoted by the Oxford English Dictionary at "Shamrock".

[27] On the importance of Easter and Saint Patrick in Ireland and on one mention of a house made of "green" wood connected with the Saint, see ARMAO Frédéric, "de Beltaine à Pâques", *Etudes Irlandaises*, Lille, 2002, II, pp. 29-41.

[28] For example in Saint Patrick's Cathedral (Dublin), Carlow Cathedral (Carlow), St. Mary's RC Church (Ballinrobe), Sacred Heart Church (Galway) to name but a few. It is true however that this propensity seems more likely to be found in the most modern representations of the Saint and is particularly popular in the American tradition.

[29] The Irish tricolours were mentioned as early as the 1830s. See source *supra.*

[30] See Article 7 of the new Constitution: "The national flag is the tricolour of green, white and orange"

> The green represents the older Gaelic tradition while the orange represents the supporters of William of Orange. The white in the centre signifies a lasting truce between the 'Orange' and the 'Green'.[31]

In other words, the Irish flag symbolizes a lasting peace between the "green" Gaelic, that is to say the Catholics, and the "orange" Protestant supporters of William – a political and religious symbol indeed. Saint Patrick, the Catholic Patron Saint of Ireland, is closely connected to the color green and the association with the color is perhaps more lively than ever today. A short study of the celebration of Saint Patrick's Day in the 20th and 21st century will show just how important (and conversely how political) this association is.

The first actual recording of the tradition of wearing shamrocks and greeneries on Saint Patrick's Day dates back to the 17th century[32] and it grew increasingly popular in Ireland, Great Britain and the United States in the 18th and 19th century. The first official Saint Patrick's parades occurred in North America even before the creation of the United States (for instance New York, 1762 and Boston, 1737)[33] and even if the Patron of Ireland was celebrated as early as the early Middle Ages in Ireland, the first official parade in Dublin only occurred in 1931. Interestingly enough, it is only in 1995 that the decision was taken to make out of this day a real cultural and therefore touristic event in the capital of the Republic of Ireland.[34] Today, the celebration of Saint Patrick's Day is a worldwide phenomenon and the color green as well as the shamrock are the emblems of Ireland throughout the world. It would obviously be impossible to mention here every single instance of that fact. In the last couple of years however, a number of significant events took (and still take) place in many countries on the 17th of March involving the use of the color green. Many cultural landmarks and historical monuments in Ireland –the first being

---

[31] http://www.taoiseach.gov.ie/eng/Youth_Zone/About_the_Constitution,_Flag,_Anthem_Harp/National_Flag.html

[32] *The Wearing of the Green, a History of Saint Patrick's Day, Op. cit.*, p.25.

[33] *Ibid.*

[34] According to the official website of the Saint Patrick's Day Festival (http://www.stpatricksfestival.ie) the first official celebration happened the following year "to offer a national festival that ranks amongst all of the greatest celebration in the world, to create energy and excitement throughout Ireland via innovation, creativity, grassroots involvement, and marketing activity, to provide the opportunity and motivation for people of Irish descent (and those who sometimes wish they were Irish) to attend and join in the imaginative and expressive celebrations, to project, internationally, an accurate image of Ireland as a creative, professional and sophisticated country with wide appeal."

Saint Patrick's Cathedral in Dublin– are illuminated in green. In the United States, the Empire State Building turns green just like the fountain of the White House and the Chicago River (starting in the 1960s) are dyed green every year in the honor of the Saint, to name but a few examples. From Australia to Canada,[35] from London to the Caribbean,[36] from Monaco, France, Italy[37] to Dubai and its famous Burj Al Arab hotel that bore for some time a gigantic green shamrock on its front, the celebration of Saint Patrick has turned into a global tradition the color of which undoubtedly remains the "green" of Saint Patrick and its emblem the three-leaf clover.

It is interesting to remind that, originally, the connection between Ireland and the color green most likely derived from the notions of Nature through the idea of a "green" Ireland which, on Saint Patrick's Day, left winter behind to fully embrace spring;[38] as early as the 18th century, the symbolism of the color shifted and "green" started to represent the need for an independent Catholic Ireland as celebrating Saint Patrick's Day while displaying green colors echoed the inherent Catholic identity of the Saint and his shamrock. It therefore became a symbol of Irish independence and freedom or, more broadly speaking, of Irishness in the most political sense of the term. Today, a new change in meaning and substance is emerging: the color green is being used and taken over by commercial and marketing ends. Saint Patrick's Day and the color of Saint Patrick's Day are becoming a means to sell and promote. Green shirts and shoes, green hats, green bagels, green pretzels, green beer, green balloons in the shape of three-leaf clovers and much more are sold throughout the world on that specific date. Interestingly enough, the digital market is not left behind and companies like Guinness –whose impact on the marketing of Saint Patrick's Day should not be neglected – released in 2011 an unmistakably "green" mobile application which promoted the festival. Similarly, the video game industry never fails to use Saint Patrick's Day's strong imagery to promote and sell new versions of their top-selling hits.

---

[35] The Opera of Sidney, the CN Tower in Toronto, the Niagara Falls were all illuminated in green.

[36] The London Eye turned green and the parade of the island of Montserrat is a peculiar (and still very popular) local tradition.

[37] The Prince's Palace of Monaco, the Leaning Tower of Pisa, the Moulin Rouge…

[38] It should be noted however, that, traditionally, spring began on Saint Brigit's Day (also called *Imbolc*, e.g. the 1st of February), especially in the most rural parts of Ireland. The view that Saint Patrick's Day is, at least partially, associated with spring is nevertheless accepted by most researchers.

To give but one example, the multi-million selling *Angry Birds* series conveniently launched in 2010 and 2011 a brand new edition pack: *Angry Birds Seasons* was offered in a variety of themes, including a Halloween theme, a Valentine's Day theme and a Saint Patrick's Day theme. That last example is all the more interesting that the subtitle of the game was: "Go green, Get lucky".[39] The ecological slogan is actually often re-used and recycled in the context of Saint Patrick's Day and anything can become a reason to "go green" especially when it is about selling a – not necessarily eco-friendly – product, be it a video game, an alcoholic beverage, a humorous T-shirt or a stuffed Leprechaun.

This "recycling" of the symbolism of the color green in Ireland is quite popular today, especially in the field of politics. Since the Irish rebellions, the color has been associated with nationalism and republicanism. This tradition is still very lively and the green of the Irish flag, as mentioned earlier, echoes the so-called Gaelic tradition when orange is a direct reference to William of Orange and the loyalists. Those colors are still used to this day by supporters of each political side. The examples are too numerous to be listed here: let's simply mention the fact that on May 17th 2011, demonstrations against the Queen's visit in Ireland were organized in the streets of Dublin; placards bore the slogan "Britain out of Ireland" where "Britain" was written with orange letters and Ireland green letters, which obviously owed nothing to chance.[40]

More broadly speaking, the color has become in the 20th century the unofficial color of the Irish republic, its visual representation, a sort of concrete metonymy of the nation and its people. The example of sport is in itself quite telling: on modern rugby or football national teams outfits, the rather discreet use of the color orange and the overwhelming color green leaves very little room for doubt: green has now become the representation, the graphic archetype of Ireland. The same could be argued about Saint Patrick and his legendary green tool, the Irish clover, especially in the political context of international relationship, and more specifically Irish-American bonds. Since 1952 tradition has held that on Saint Patrick's Day the Irish prime minister or the Irish ambassador to the United States hand shamrocks to the American President, regardless of

---

[39] http://www.angrybirdsnest.com/angry-birds-seasons-go-green-get-lucky-update-is-here/

[40] See the article of Henry McDonald "Queen's visit to Ireland: Garda foil plans to disrupt Dublin events" in *the Guardian*, Tuesday 17 May 2011 (http://www.guardian.co.uk/uk/2011/may/17/queens-visit-ireland-garda-foil-disruption).

their political affiliation.[41] When Barack Obama has posters and signs created to promote his election or re-election to the Irish-American minority, he uses and abuses of the color green and the shamrock while amusingly changing his name to O'Bama, reminding the voters of his Irish ancestry.[42] The shamrock and its color are therefore political and diplomatic vectors or symbols of Ireland. When Queen Elizabeth decides to go visit the Republic of Ireland for the first time, she naturally and quite diplomatically chooses to wear green clothes (yet featuring a slight touch of blue). Stewart Marvin, one of the fashion designers of the Queen, clearly states that the choice in the colors of the monarch's apparel is never a matter of coincidence when she visits a foreign country[43] and that was certainly not the case there; once again, the "Go Green" play-on words was used by some journalists and the title "The Queen Goes Green in Ireland" was unoriginally – yet quite appropriately – chosen by People Magazine.[44] The same happened when on 17 March 2012 Kate Middleton chose to visit the Irish Guard: the Duchess of Cambridge presented sprigs of shamrocks to the guards – conforming to a tradition dating back to 1901– while wearing a deep green coat dress and a gold shamrock brooch.[45]

If most Irish political parties[46] use logos where the color green is predominant, with the notable exception of *Fine Gael* and some left-wing and workers' parties,[47] the political use of the color is most notable as regards ecology. It is quite clear that everything prompts the Irish ecologists to comply with the unspoken rules of this green political

---

[41] *Fina Gael*'s Enda Kenny in 2012, *Fianna Fáil*'s Brian Cowen before him to name but the last instances of this tradition.

[42] See the article "Obama Campaign Touts his Irish Heritage" by Byron Tau in *Politico,* 17 February 2012 (http://www.politico.com/politico44/2012/02/ obama-campaign-touts-his-irish-heritage-114862.html).

[43] For the interview of Stewart Marvin, see "Des Racines et des Ailes", broadcast 9 mai 2012 on France3, hosted by Louis Laforge. Report by Hélène Frandon, Antoine Placier, Laura Kalmus, Didier Pêcher, Pascal Montagna, Jérôme Sarfati, Marc Mazeas.

[44] For a picture of the Queen and the article of People Magazine, visit http://www.people.com/people/package/article/0,,20395222_20490446,00.html

[45] See the article "Green goddess: Duchess of Cambridge presents shamrocks to Irish Guards at St Patrick's Day parade" by Natalie Evans in *Mirror,* 17 March 2012 (http://www.mirror.co.uk/news/uk-news/duchess-of-cambridge-presents-shamrocks-to-irish-764890).

[46] Fianna Fáil, Sinn Féin, Éirígí and the Republican Sinn Féin.

[47] The Workers' Party, the Socialist Party and the Labour Party still favor the color red for obvious historical reasons.

marketing: green is the unofficial color of Ireland just like it is the obvious color of ecology; under these circumstances, why would the official logo of the Irish ecologists not be green? Today the logo of the *Comhaontas Glas*, the Irish Green Party, features a stylized sunflower with yellow petals and a light-green center on a dark-green background, which should not be surprising. However, it must be noted that the Green Party in Northern Ireland, the Northern Ireland region of the *Comhaontas Glas* created in 1983, has two official colors: green and blue. The logo of the Green Party in Northern Ireland is an adapted version of the standard green logo of the Green Party: the sunflower is now presented on a deep blue background. Politics and national identity are once again symbolically united as the choice of those two colors cannot be accidental. The color blue was most likely chosen for historical reasons; obviously, it cannot be understood as a provocation towards the Republic of Ireland (as the Green Party of Northern Ireland is an offspring of the Irish Green Party): it must be considered as an example of the symbolic complementarity of those two colors. If the color used by the ecologists from Northern Ireland obviously reminds us of the color chose by Henry VIII to represent the island thus placed under his authority, it must most of all be associated to Saint Patrick's blue and therefore be considered as an alternative way to represent Ireland – as the self-professed cross-community nature of the party prevents it from any Unionist or Loyalist endeavors.

The question of whether the use of the color green by Irish ecologists in the context of graphic design is supposed to embody the green of ecology or that of Ireland, in other words the "green of Nature" or the "green of Identity", is difficult to answer. The truth probably lies somewhere in between as we should remember that the symbolic use of that color in Ireland probably comes from the flag of the Irish rebellion but also from two "natural" symbols: the bough of liberty and the legendary clover of Saint Patrick. Finally, the fact that Ireland literally looks green must not be minimized. It is therefore possible to briefly sum-up the history of the color in Ireland: originally, or more precisely "as far back as we can trace", the color seems to have been associated to nature, to "greenery" or, in other words, to the "natural" aspect of Ireland. The color was then Christianized through the prism of Saint Patrick and his legendary life: his hagiographers and supporters associated the color to religion as the green shamrock was supposed to have epitomized the concept of Trinity. The color was then politicized: Saint Patrick is obviously linked to Catholicism and green became the symbol of the struggle against the English settlers in the 17th and 18th centuries. Finally, in more recent times – i.e. from the second half of the 20th century on – the Irish green took its current meaning under the double influence of the marketing of Saint Patrick's Day on the one hand and the political ecology on the other. Nature, religion, politics, ecology, marketing: all those elements combined and gave birth to a true national – yet non official– symbol: the Irish Green. The past of Ireland has been reused, adapted, recycled by the religious, the political, the advertising man or simply the average citizen so that this past has finally become itself a vector of creation –or rather re-invention– of identity. The myths of Nature, the ecologic mythology of Ireland have actually led for over 1,500 years to a constant regeneration of Irishness that is, in the true sense of the word, to a persistent recycling of identity.

Bibliography

Armao, Frédéric, "De Beltaine à Pâques", *Etudes Irlandaises*, Lille, 2002, II, pp. 29-41.

Brennan, Paul et Peyronel, Valérie, *Civilisation Irlandaise*, Paris, Hachette Supérieur, 1999.

Caulfield, James, The Great Irish Rebellion, Secret and Silent Men of 1798, Victoria, Trafford, 2004, p. 114.

Cronin, Mike & Adair, Daryl. *The Wearing of the Green, a History of Saint Patrick's Day,* London, Routledge, 2006.

Guillaume, André, *L'Irlande, une ou deux nations ?*, Paris, P.U.F., 1987.

Hayes-McCaoy, G.A., A History of Irish Flags from the earliest times, Boston, G.K.Hall&Co, 1979.

Kee, Robert, *The Green Flag, A History of Irish Nationalism,* London, Penguin Books, 2000 (1972).

Mackillop, James, *Dictionary of Celtic Mythology,* Oxford, Oxford University Press, 1998

Maignant, Catherine, Histoire et Civilisation de l'Irlande, Paris, Nathan, 1996

Mollard-Desfour, Annie, *Le vert : Dictionnaire de la couleur, mots et expressions d'aujourd'hui XXe-XXIe*, Paris, Editions du C.N.R.S., 2012.

O'Kane, Michael, *Considering the Irish Greens, An Ethnographic Approach to Identity and Environmentalism,* Verlag DM, 2009.

O'Laughlin Michael C., The Irish Book of Arms, Irish Genealogical Association, 2001

Pakenham, Thomas, The Year of Liberty. The Great Irish Rebellion of 1798, London, Hodder & Stoughton, 1969.

Trinquier, Jean, "Les Vertus Magiques et Hygiéniques du Vert dans l'Antiquité", in *Couleurs et Visions dans l'Antiquité Classique,* Rouen, Publications de l'Université de Rouen, 2002, pp. 97-128.

Zimmermann, Georges Denis, Songs of Irish rebellion: Irish political street ballads and rebel songs, 1780–1900, Dublin, Four Courts Press

PART VI:

CLIMATE CHANGE ISSUES

THE STERN REVIEW (2006): ITS PLACE IN BRITAIN'S AND THE WORLD'S ASSESSMENT OF THE ECONOMIC COSTS OF GLOBAL WARMING AND THE COSTS OF COMBATING CLIMATE CHANGE

NICHOLAS SOWELS
UNIVERSITY OF PARIS 1 PANTHEON-SORBONNE

The Stern Review stands out as a major, comprehensive study of the *economic* impacts of climate change. Compiled by a team of economists and researchers under the direction of Sir (now Lord) Nicholas Stern, the government's chief economic advisor at the time, the review provides wide-ranging analysis of likely climate change, the costs which such climate change is estimated incur to the world economy if no particular action is taken to limit change, and the costs of seeking to mitigate such climate change.

This chapter sets out to present succinctly the findings of the Review, the underlining economics used to reach these findings, to examine the Review's reception and look at its impact on policy in the United Kingdom.

Remit, findings and recommendations

In July 2005, the then Chancellor of the Exchequer, Gordon Brown, asked Sir Nicholas Stern to lead a review on the economics of climate change. The terms of reference of the review included, "examin[ing] the evidence on:

- The implications for energy demand and emissions of the prospects for economic growth over the coming decades [...];
- The economic, social and environmental consequences of climate change in both developed and developing countries;

- The costs and benefits of actions to reduce the net global balance of greenhouse gas emissions from energy use and other sources, [...], taking into account the potential impact of technological advances [...]; and
- The impact and effectiveness of national and international policies and arrangements in reducing net emissions in a cost-effective way and promoting a dynamic, equitable and sustainable global economy [...] (HM Treasury/Cabinet Office, 2005).

After consultation with key stakeholders, internationally and domestically, the review also had to "provide:

- An assessment of the economics of moving to a low-carbon global economy [...].
- An assessment of the potential of different approaches for adaptation to changes in the climate (ibid).

Specific attention was to be given to the United Kingdom, to assess how this analysis applies to the case of the UK, in the context of its existing climate change goals.

The Review starts by examining the scientific evidence concerning climate change, drawing on the Third Assessment Report (TAR) of the Intergovernmental Panel on Climate Change (IPCC), published in 2001. It notes that the current (2006) level of greenhouse gases (GHGs) in the atmosphere is equivalent to about 430 parts per million (ppm) CO_2equivalent (CO_2e), compared with only 280 ppm before the Industrial Revolution, with annual increases of around 2.3 ppm.

Significantly, the Review notes that even if annual emissions do not increase, then it is estimated that the stock of GHGs is likely to rise to 550pm CO_2e by 2050. If this concentration is then maintained, global temperatures are estimated to rise by 2-5°C or even higher: based on TAR and other forecasts. However, if emissions increase, especially as fast-growing countries invest in "highcarbon infrastructure" and as the use of energy expands globally in transport, then the 550ppm level (i.e. twice the level before industrialisation) could be reached by as early as 2035.

If annual emissions continue until the end of the 21st century at their current rate, then CO_2e levels could rise to between 650-1200 ppm by 2100, leading to global warming of 3-10°C or more, even assuming that no additive feedback takes place (Stern, 2007, p15-16). To put this in other words, under the Review's "Business as Usual" (BAU) scenario, the stock of greenhouse gases in the atmosphere could more than treble by the end of the 21st century, leading to at least a 50% risk of the global average temperature rising by at least 5% (Stern, 2007).

In assessing the economic costs related to climate change, the Review adopts three approaches: i) it examines the physical impacts of climate change on economic activity, human life and the environment; ii) the Review models the costs of continuing with BAU, and iii) it undertakes a cost-benefit analysis of measures to limit climate change.

On the assumption that global temperatures rise by 2-3°C over the next fifty years, the Review sets out likely impacts on human activity and the environment. These impacts include: melting glaciers leading to flooding and falls in water supplies which menace one-sixth of the world's population; falling crop yields, especially in Africa; more deaths in higher latitudes due to malnutrition, heat stress and diseases (like malaria); rising sea levels leading to flooding of tens to hundreds of millions of people each year (with warming of 3-4°C); change to ecosystems as around 15-40% of species face extinction after only a 2°C rise in temperature.

Significantly, these effects will not be evenly distributed, though emissions anywhere in the world have a global impact. The poorest countries and people are set to suffer earliest and most.

The second approach by the Stern Review involves using an integrated assessment model (IAM). Having considered that existing economic models forecasting the monetary cost of climate change fail to capture fully the likely cost of climate change, the Stern Review uses a model (PAGE2002) which can take account of a large range of risks, including possibilities of a "high climate" scenario which was designed to explore the consequences of higher temperature increases. This could result from stronger feedback effects, notably weakened carbon sinks (as plant and soil respiration increases, offsetting sink capacity), and from higher natural methane releases (currently held in wetlands and permafrost, but released at higher temperatures).

On the basis of this modelling, the Review concludes that with no change in economic practices – i.e. if the world carries on with business-as-usual (BAU) – then the average reduction in per-capita consumption is estimated to be 5% "now and forever". Yet, this is only the lower threshold. If direct, so-called "non-market" impacts on the environment and human health are taken into account, then the costs in global per-capita consumption rise from 5% to 11%. With feedback effects, the possible costs of climate change could rise from 5% to 7%, and from 11% to 14%, taking into account non-market effects. Also, the effects of climate change are set to affect the world's poorer regions disproportionately. If these effects are weighted, then the costs of climate change in the region of 5-

6°C could be higher by a quarter. This leads to a total reduction in consumption per head of 20%.

The Report goes on to evaluate the costs of cutting emissions. It notes that up until today, 70% of all CO_2 emissions have been produced in North America and Europe, but that in the future most emissions growth will occur in developing countries. Cutting emissions with the aim of stabilising greenhouse gas levels in the atmosphere at 500 to 550 ppm CO_2e will entail significant costs, rising to a level of about 1% of global GDP by 2050. The Report notes that emissions can be cut in four main ways, with different costs. These include: i) reducing demand for goods and services which produce a lot of emissions; ii) increasing energy efficiency; iii) action on non-energy emissions, such as avoiding deforestation; and iv) shifting to lower-carbon technologies for power, heat and transport. According to the macroeconomic modelling approach used by the Review, while the average cost for stabilisation may be 1% of GDP, there could actually be a range of costs, stretching from -2% (i.e. an addition to GDP) to 5% (a considerably higher cost). Much depends on the pace of technological innovation and the efficiency with which policy is applied across the world. The Stern Review states quite clearly, however, that delaying action carries a particularly high price.

The third approach used to examine the costs and benefits of action on climate change compares the marginal cost of reducing emissions and the social cost of carbon (i.e. the costs to society of emitting carbon dioxide which are not directly incurred by the emitter). On a business-as-usual basis, the Review estimates the social cost of to be $85 per tonne of CO_2. The Review notes that this cost estimate is higher than the figures usually given in other evaluations, and stresses too that such a cost is considerably greater than the costs of reducing emissions: the abatement costs. The Report calculates that strong mitigation policies taken today (i.e. in 2006), would lead to net benefits over time in the order of $2.5 trillion.

To reduce emissions and mitigate the effects of global warming, the Review puts forward three essential policies: i) carbon pricing, ii) technology, and iii) the removal of barriers to behavioural change.

This primary emphasis on carbon pricing stems from the fact that the Review presents greenhouse gas emissions as being an "externality", in economic terms. In other words, producers of greenhouse gases do not pay the full costs of what they are doing: the overall costs to the environment are not directly included in the costs which greenhouse gas emission activities face. Instead, they are off-loaded on world's society today and on future generations. To remedy this, policies have to be adopted which bring the private cost of creating emissions in line with their social cost.

This essentially means raising the private cost of emissions to discourage them.

The Review points out that there are several ways of doing this, combinations of which governments are likely to adopt. Governments can directly tax emissions, which has the advantage of discouraging such activities, while at the same time raising revenues. The latter could be used to finance emissions' reducing research and/or dealing with the consequences of global warming. Direct government regulation is also a means for intervening literally, directly to limit the output of greenhouse gases.

The Review however leans more towards carbon pricing. By fixing, ideally, a global carbon price, and creating markets in which greenhouse gas emissions can be traded, the market mechanism can be used to shift activity in the most efficient way. Higher prices and trading should have the effect of discouraging emissions more rapidly in those areas where they can be cut easily, in particular, while allowing other activities to continue, albeit facing higher carbon prices. In principle, this should help bring about greater cuts with the least disruption to economic activity.

The Review links technology change inextricably to the issue of pricing. It states quite unambiguously that halting the rise of carbon emissions will require the implementation of a "wide range" of low carbon technologies. The Review declares that this entails significant private sector involvement, in both R&D and the production of such technologies. At the same time, government action is important in establishing the necessary incentives for private actors to develop low carbon technologies. More specifically, government has important roles to play in shaping price incentives for firms to invest in low-carbon technology, which is currently usually more expensive than fossil-fuel technology, and in supporting fundamental research. The case for support of new forms of electricity generation is especially strong.

Elsewhere, government intervention is required in stimulating behavioural change in agents. For the Stern Review, this relates to questions concerning the information available to agents, the transaction costs they face and their organisational practices.

A crucial, if not *the* fundamental component of meeting global warming lies in international cooperation. This however faces substantial hurdles. As N. Stern has clearly noted elsewhere, in the case of global warming as an externality:

> ...we have many jurisdictions, weak representation of those most affected (future generations), long-term horizons, a global scale, major uncertainties, and important interactions with other market failures. Thus,

whilst the standard theory can provide useful initial insights, we have a much deeper and more complex economic policy problem. We have a problem of intertemporal international collective action with major uncertainty and linked market failure" (Stern, 2006).

To deal with this situation, the Review calls for a clear global perspective concerning the urgency of the problems to be addressed, drawing on multinational frameworks and coordinated action. The UN Framework Convention on Climate Change (UNFCCC) with its Kyoto Protocol and other informal dialogues constitute a starting point for action. But more obviously needs to be done, and quickly. Specifically, the Review calls for a "broadly similar" carbon price to help limit the overall costs of reductions, and technologies should be transferred to developing countries.

This in turn raises the issue of transferring resources to developing countries. The Review estimates that the costs of pursuing low-carbon investments will run to \$20-30 billion per year, and that this will require a substantial increase in the flow of funds from carbon trading systems in industrialised countries – like the EU ETS – to the developing world. Existing measures (like the Clean Development Mechanism) and existing institutions (the international financial institutions) will have important roles to play in this process.

Reception of the Stern Review

The reception of the Review was, not surprisingly, varied. The Labour government strongly endorsed it. Tony Blair, the Prime Minister, embraced the Review's findings and recommendations wholeheartedly, noting that the consequences of inaction were "literally disastrous", not just in the distant future but "in our lifetime". As a result, he called for action to be taken immediately, stating that it was simply not possible to wait for lengthy negotiations such as those which had preceded Kyoto (BBC, 31 Oct 2006).

Many leading economists similarly welcomed the Review. It was openly supported by several Nobel Laureates (James Mirrlees, Amartya Sen, Joseph Stiglitz and Robert Solow) and others, as providing a comprehensive, serious analysis of the issue of climate change and possible policy responses to it. To quote J. Stiglitz, for example, in the introductory pages of the subsequent publication of the Review:

The *Stern Review* of *The Economics of Climate Change* provides the most thorough and rigorous analysis to date of the costs and risks of climate

> change, and the costs and risks of reducing emissions. It makes clear that the question is not whether we can afford to act, but whether we can afford not to act. .. . And it provides a comprehensive agenda – one which is economically and politically feasible – behind which the entire world can unite in addressing the most important threat to our future well being (Stiglitz in Stern, p ii).

Nor was support confined to progressive economists such as the above. Paul Wolfowitz, the World Bank's President at the time and former neo-conservative Deputy Secretary of Defence of the Bush Administration, also broadly backed the Review, as did, for example, Adair Turner, former Director of Britain's Confederation of British Industry.

Others contested the study. They included the sceptics concerning climate change, like Ruth Lea, director of the Centre for Policy Studies, the think-tank set up in the 1970s by Keith Joseph and the "New Right" close to Margaret Thatcher. In an op-ed article in *The Daily Telegraph*, she began by challenging the science of climate change as not being necessarily correct, while asserting that "the climate system is far too complex for modest reductions in one of the thousands of factors involved in climate change (i.e. carbon emissions) to have a predictable effect in magnitude, or even direction". More generally, she questioned the rational of Britain leading international policy as this would not only likely be ineffective but weighed down on British competitiveness. In short, she concluded that the Stern Review would most likely be used by the Labour Government as an excuse to raise taxes on fossil fuels (Lea, R., 2006).

In a similar vein, former Conservative Chancellor, Lord (Nigel) Lawson, criticised the Review before the House of Commons Treasury Select Committee and elsewhere. He stated his scepticism notably of persuading China to cut back on it carbon emissions. More generally, Lord Lawson called into question the objectivity of the Review, claiming that it was in fact a "work of advocacy" produced by Nicholas Stern on behalf of the Labour government, and in line with the government's policy orientations. He also criticised the Review for exaggerating the costs of climate change, while understating the benefits of warming, and the costs of mitigation (HCTSC, 2008, pp 11-12).

This last point relates directly to the more technical, yet fundamental criticism levelled at the Review, namely the implied discount rate it uses to calculate the future costs and benefits of climate change and mitigation policies, at present values. In the words of William Nordhaus of Yale University who criticised the Stern Review on precisely this point, the discount rate may seem to be an "arcane" issue but it has radical consequences for the economics of climate change. A low discount rate raises the future costs of climate change in current terms, and therefore provides strong accounting support for enhanced and more costly actions in the present to mitigate climate change and offset these costs. Alternatively, a higher discount rate greatly changes the economics of global warming. By decreasing the future value of the costs of climate change relative to the present costs of mitigation, a higher discount rate has significant policy implications in the near- to medium-term: put simply it raises the question of why present generations should make important efforts now to reduce global warming, with all the consequences this will likely to entail for current living standards, when future generations are likely to be able to deal much more easily with the consequences of global warming, given their far higher income levels (Nordhaus, 2006).

More specifically, according to W. Nordhaus, the Stern Review breaks with previous economic analysis and hence policy prescriptions by assuming that the discount rate is nearly zero. Previous economic estimations, using more or less the same evidence on climate change as Stern, had usually assumed higher discount rates. As a result, their policy prescriptions were based much more on addressing global warming progressively at first, and proposing greater actions to mitigate climate change in the future, as it occurs. The result is a "climate policy ramp", in which policies expand over time. W. Nordhaus states clearly that the logic behind such a policy ramp is that future opportunities – in a richer global society – will open up new investment opportunities to tackle climate change, with low-emissions technologies, etc. For William Nordhaus, the Stern Review's conclusions and policy prescriptions do not hold up, as they are largely invalidated by changes in assumptions concerning the discount rate.

A similar critique is also put forward by Martin Weitzman, a specialist both in the economics of resource use and the environment at Harvard University. M. Weitzman too notes that the assumptions on the discount rate found in the Review are easy to challenge, and he criticises it for providing a one-sided analysis based on this. However, he ultimately comes out broadly supporting the Review, but by taking a significantly different approach. More specifically, Weitzman notes that the issues of climate change and mitigation policies should not be viewed in terms of consumption smoothing: i.e. essentially reducing consumption now to pay more for mitigation and as a result be able to consume relatively more in the future, given that the impacts of global warming have to some extent been controlled. Instead, Martin Weitzman notes that most economic analyses take little account of extreme, but improbable outcomes linked to global warming. He states that these need to be researched further, and that policy on climate change should also be seen as "insuring" against such outcomes. From this point of view, M. Weitzman's overall conclusion on the Stern Review is more favourable, as the immediate action it proposes should be viewed as a precautionary insurance policy against extreme, if somewhat less likely risks.

As for Nicholas Stern, he addressed the issue of discounting both in the Review, and later before the Treasury Select Committee. His arguments in both cases were essentially twofold. On the one hand, he stressed that the Review adopted usual accounting approaches used by the Treasury in evaluating infrastructural investments and employed more generally in economics in such circumstances. Secondly, he justified the use of a low discount rate by the ethical argument that the well-being of future generations – even though, or perhaps precisely because – they are not represented in present policy discussions should be given equal importance to today's generations (HCTSC, pp 14-15).

Furthermore, a mere two years after the Review was published, Lord Stern went on the record stating that climate change was occurring faster than had originally been estimated in the Review. Speaking in June 2008, N. Stern came out in support of holding CO2 levels below 500ppm – the middle value of the range announced in 2006. However, to achieve this, he estimated that the new mitigation effort would be equivalent to 2% of global GDP: a notable increase in costs (*The Guardian*, 26 June 2008).

The Impact of the Stern Review on UK Government Policy

Not surprisingly, in the immediate aftermath of the publication of the Review, Gordon Brown as Chancellor of the Exchequer announced a series of initiatives and aspirations combining many of Review's approaches. The UK governments' policies had already been moving in this direction, shown for example in the ratification of the Kyoto Protocol and moves to cut carbon emissions with respect to the 1990 levels. But the Review clearly provided additional impetus to policy. In the Pre-Budget Report Statement made to the House of Commons in December 2006, G. Brown declared that "partnership of responsible individuals, companies and governments [was] vital to meeting the environmental challenge". He went on to list the Government's environmental priorities as follows:

- To support market mechanisms and incentives to promote carbon trading;

- To encourage the development of London as the "world's leading centre for carbon trading", in collaboration with "major financial institutions;
- To develop biofuels in partnership with Brazil, Mozambique and South Africa;
- To preserve rainforest by "working with Latin American and Asian countries;
- To foster in partnership "clean coal with China and India;
- To launch a feasibility study with Norway for the capture of carbon dioxide and its storage under the North Sea;
- To reduce carbon emissions from homes, which account for 30% of the total, by ensuring that new homes are carbon free "within years", a policy supported for a limited time by exempting these homes from stamp duty;
- To encourage energy audit in existing homes (HM Treasury, 2006).

In the subsequent Budget (HM Treasury, 21 March 2007), the Chancellor went on to announce that the Government had "secured support" for strengthened European carbon trading, and had agreed to the European scheme for reducing emissions by 20% by 2020 (and potentially 30%). In fact, this was part of a broader package formulated by the EU in 2007 and 2008, and which also refers to the Stern Review. Indeed, the EU has adopted its so-called 20/20/20 targets, entailing cuts in emissions by 20%, raising renewable energy supplies to 20% and implementing energy savings of 20%, by 2020 (European Commission, 2008).

Gordon Brown also stated that Britain would begin developing carbon capture technology (i.e. technology which allows carbon dioxide produced notably in electricity generation to be captured and stored underground). He also reiterated the Government's commitment: to supporting clean coal initiatives with India and China; to providing finance for Africa to prevent rainforest destruction; and in particular the Government's policies for reducing emissions from Britain's housing sector, etc. (HM Treasury, 2007).

In 2008 Budget, the Chancellor took these objectives further. In particular, he announced that new non-domestic buildings should be zero-carbon from 2019 onwards and that the Climate Charge levy (introduced in 2000) would rise in line with inflation. Gordon Brown also set out the need to reduce carbon emissions from cars, noting that Britain's 30 million cars, vans and lorries were responsible for 22% of the country's emissions. To help reduce such emissions, he stated the Government's support for bringing down the EU cap on emissions in 2020 from 130g per km to

100g. In addition, he announced future changes to Vehicle Excise Duty to make it less expensive to drive low emission cars compared to high-emission ones (HM Treasury, 2008).

More generally, the Labour Government headed by Gordon Brown, who became Prime Minister in June 2007, passed the Climate Change Act in 2008. It sets out a legally binding obligation to "ensure [that the] net UK carbon account for the year 2050 is at least 80% lower than the 1990 baseline", which covers reductions in CO2 and other greenhouse gases: the target for carbon emissions by 2020 is "at least 26%", later revised to 34%. The Act introduced five-yearly carbon budgets, covering 2008-2012, 2013-17 and 2018-22 to help target the reduction in emissions. The Act also covers numerous detailed issues relating to these reductions, including questions concerning the inclusion of emissions from international transport to and from the UK, the limits of buying up CO2 emission rights from abroad, and powers to develop domestic trading markets more quickly using secondary legislation, measures on biofuels, household waste, etc. In addition, the Act created the Committee on Climate Change, whose functions include notably providing advice and assessment of meeting the carbon targets set out in the five-yearly carbon budgets (Climate Change Act, 2008). The Labour Government also created the Department of Energy and Climate Change (DECC) in the autumn of 2008, with responsibility for overseeing government policy on climate change.

The broad message of the Stern Review as well as the emission targets set by the Act for 2020 and 2050 have been adopted by the present coalition government, that took office in May 2010 and which is headed by the Conservative Prime Minister David Cameron. Accordingly, in May 2011, the Energy and Climate Change Secretary Chris Huhne set out the Fourth Carbon Budget for 2023 to 2027, in line with the recommendations of the Climate Change Committee. It legally fixes Britain's GHG emissions target for the period at 1950 million tonnes of CO2 equivalent (MtCO2e). This constitutes a 50% reduction on the 1990 baseline, and is consistent with the 80% target for 2050 (DECC, 17 May, 2011).

More substantively, in its White Paper on "Electricity Market Reform" (EMR) published in July 2011, the DECC outlines the significant challenges faced by the UK electricity industry in meeting CO2 reduction targets. The White Paper specifically makes reference to the Stern Review noting (somewhat selectively): " [t]he review concluded that a transparent and predictable carbon price is the most cost-effective way to encourage emitters to invest in alternative low carbon technologies and change consumer spending patterns" (DECC, July 2011, p 27). Concretely, the

White Paper states bluntly that for the UK to reach its 2050 target, electricity generation will have to be "largely decarbonised by the 2030s". Given that the UK is also set to lose about "a quarter (around 20 GW) [its] existing generation capacity as old or more polluting plants close" the challenge to invest in new, lower-emitting power generation and renewable electricity generation is huge: up to £110 billion is estimated to be required in investment in electricity generation and transmission by 2020, more than twice the existing investment rate (ibid. pp 5-6).

The White Paper states that the existing market mechanisms are not sufficiently stable and predictable to nurture long term investment. It therefore proposes several policies for reforming the market, and achieving an environment conducive to stable contracts. These include:

- long-term contracts for both low-carbon energy and capacity;
- institutional arrangements to support this contracting approach;
- continued grandfathering, supporting the principle of no retrospective change to low-carbon policy incentives, within a clear and rational planning cycle; and
- ensuring a liquid market that allows existing energy companies and new entrants to compete on fair terms" (ibid, p 7).

These are supplemented by two further measures which involve:

- the introduction of a Carbon Price Floor (CPF) to reduce uncertainty, put a fair price on carbon and provide a stronger incentive to invest in low-carbon generation now...
- an Emissions Performance Standard (EPS) set as an annual limit equivalent to 450g CO2/kWh at baseload to provide a clear regulatory signal on the amount of carbon new fossil-fuel power stations can emit... (ibid. p 7).

In contrast to this stark outlook, the DECC issued a more upbeat report on reducing emissions at the end of 2011. *The Carbon Plan* too restates the objectives of cutting carbon emissions by 80% by 2050, with the immediate target of 34% by 2020. It notes that so far the UK has scored some successes in achieving these targets, notably as the "dash for gas" in electricity generation (in the wake of electricity privatisation in 1990) helped reduce emissions from electricity generation by a quarter, between 1990 and 2010. Emissions from buildings have also fallen by 18%, thanks to the use of condensing boilers and insulation in cavity walls which now covers 60% homes with such walls. Significant reductions in emissions have also been recorded in industry (down by 46%, though output has increased by an average 1%), and agriculture (a third less emissions, brought on largely by more efficient procedures and less emissions from landfill waste). On the other hand, emissions from transport at the end of 2010 were roughly the same as in 1990, with growth recorded through to the end of 2007, followed by a decline linked to the recession (DECC, December 2011).

The Carbon Plan also notes, however, that much of the progress so far has been relatively easy, while in the future:

> ...if we are to cut emissions by 80% by 2050, there will have to be major changes in how we use and generate energy. Energy efficiency will have to increase dramatically across all sectors. The oil and gas used to drive cars, heat buildings and power industry will, in large part, need to be replaced by electricity, sustainable bioenergy, or hydrogen. Electricity will need to be decarbonised through renewable and nuclear power, and the use of carbon capture and storage (CCS). The electricity grid will be larger and smarter at balancing demand and supply" (ibid, p 4).

To achieve the objectives and these changes in technology, *The Carbon Plan* underlines the importance of developing a "portfolio of technologies for each sector", with the aim of reducing risks associated with single technologies, as well as stimulating competition between them.

> In electricity, the three parts to our portfolio are renewable power, nuclear power, and coal- and gas-fired power stations fitted with carbon capture and storage. In transport, ultra-low emission vehicles including fully electric, plug-in hybrid, and fuel cell powered cars are being developed. In buildings, the technologies will include air- or ground-source heat pumps, and using heat from power stations. Both of these are solutions proven by their use in other countries (ibid, p5).

Overall, the *Plan* is pro-active and optimistic. It notes that numerous technologies have already been tested elsewhere, and that other countries are embarking on similar commitments. At the end of the document – Annex C – the *Plan* sets out 10 pages of specific steps, measures and proposals with "start dates" and "end dates" of actions to be undertaken before the end of Parliament (i.e. the next elections which must take place in May 2015 at the latest).

In contrast, a more muted assessment of progress on achieving carbon reductions has been given more recently by the Climate Change Committee (CCC). In its Third Progress Report to Parliament (published June 2012), the CCC records that there was a 7% fall in emissions in 2011.

But this substantial reduction was very much due to a milder winter, more costly energy discouraging consumption and the effects of the recession. Emissions merely fell by 0.8% as a result of carbon lowering measures, a rate which is only a quarter of what is needed to meet carbon cutting targets (Climate Change Committee, 2012)! The report generally notes that more needs to be achieved in all sectors.

At the European level, recent developments in the CO2 emissions market also raise questions about its present ability to send the right price signals to market agents, especially in the electricity generating sector. Since early 2011, the emissions price has collapsed, falling to €7-8 per tonne, or about half its previous level, due mainly to the effects of the recession, and possibly the mild winter of 2010/2011 (Cornot-Gandolphe, p 16). This has helped stimulate the subsequent use of coal in electricity production in Europe, especially as coal prices have weakened too relative to gas. These may of course turn out to be short-term phenomena that will be reversed by future increases in the price of CO2 emissions, in the European market. But countries like Germany, which has large brown coal (lignite) reserves or Poland could well be faced with price signals that do not discourage the use of coal, which is a high CO2-emitting primary energy source. The functioning of the market mechanism for trading CO2 emission rights does not therefore seem to be completely self-evident.

The Stern Review stands out as an all-encompassing analysis of global warming, its associated costs and the costs of mitigation. It unambiguously made the case for early, strong action both by the UK government and the international community as a whole.

When it was published, it was well-received by prominent, internationally renowned economists. At the same time, some leading environmental economists, as well as sceptics of climate change, criticised the report. In particular, its implied cost discounting was seen to give too much emphasis on the (very) long term effects of climate change, hence justifying relatively costly mitigation action beginning in the present.

Broadly speaking, the Review has strengthened the case for public action in the United Kingdom to reduce CO2 emissions, a policy which had already been adopted in the wake of the ratification of the Kyoto Protocol. The Review had a direct impact on the Climate Change Act which the Labour government passed in 2008, and which sets legally binding emission reduction targets for the United Kingdom of 34% by 2020 and 80% by 2050 (compared to 1990 levels).

These targets have been accepted by the Conservative-Liberal coalition, which came to office in 2010, and which has adopted detailed, explicit policy plans for reducing emissions in electricity generation and throughout the economy.

Recent figures (for 2011) show that the UK is "on target" in bringing down CO2 emissions. However, cuts so far do appear to have been relatively easy to achieve, compared with what will need to be done in the future. Also, temporary conditions, such as the economic recession have helped reduce emissions. It therefore remains to be seen, whether the UK will achieve its emission objectives for 2020 and beyond, without profounder changes to the structure of its economy and functioning of its society.

Bibliography

BBC, "Climate change fight 'can't wait'", 31 October 2006, available at: <http://news.bbc.co.uk/2/hi/business/6096084.stm>, viewed 10/10/2012.

Climate Change Act, 2008.

Climate Change Committee, *Meeting Carbon Budgets – 2012 Progress Report to Parliament*, June 2012.

Cornot-Gandolphe, S., *The European Coal Market: Will Coal Survive the EC's Energy and Climate Policies?*, IFRI, Paris, 2012.

DECC, "UK proposes fourth carbon budget", Press Release, 11/041, 17 May 2011.

—. Planning our electric future: a White Paper for secure, affordable and low-carbon electricity, CM 8099, July 2011.

DECC and HM Government, *The Carbon Plan*, December 2011.

European Commission/Commission of the European Communities, *20 20 by 2020: Europe's climate change opportunity*, communication COM (2008) 30 final, 23/01/2008.

Guardian (The), "Cost of tackling global climate change has doubled, warns Stern", by J. Jowit and P.Wintour, 26 June 2008.

HCTSC (House of Commons, Treasury Select Committee), *Climate Change and the Stern Review: the Implications for Treasury Policy*, Fourth Report of Session 2007-08, 15 January 2008.

HM Treasury / Cabinet Office, Press release, 85/05, "Stern Review on Economics of Climate Change: Terms of reference and call for evidence announced", 12 October 2005.

—. *2006 Pre-Budget Report*, 6 December 2006.

—. *2007 Budget Statement*, 21 March 2007.

—. *2008 Budget Statement*, 12 March 2007.

Lea, R., "Just another excuse for higher taxes", *The Daily Telegraph*, 31 October 2006.

Nordhaus, W. D., *The "Stern Review" on the economics of climate change*, Working Paper 12741, National Bureau of Economic Research, December 2006.

Stern, N. "What is the Economics of Climate Change", *World Economics*, Vol 7, No 2, April-June 2006.

—. *The Economics of Climate Change: the Stern Review*, CUP, Cambridge, 2007. See also the Executive Summary (long format), at: <http://www.hm-treasury.gov.uk/media/4/3/Executive_Summary.pdf>, consulted 10/10/2012.

GLOBAL WARMING: BETWEEN DOUBTS AND CERTAINTY

PATRICK MENNETEAU

UNIVERSITY OF TOULON, BABEL RESEARCH LABORATORY & CERCLE FRANCOPHONE DE RECHERCHE ET D'INFORMATION C.G. JUNG IN PARIS

The global warming issue revolves mainly around two central questions: is it real? And, if so, what causes it? The first question was still a mooted point until a few years ago, with some scientists, such as Claude Allègre in France, claiming that rising temperatures in certain areas of the globe were made up for by falling temperatures elsewhere, and that on average, the global situation was stable. Such a position has become more difficult to maintain in face of the rising evidence concerning the meltdown of glaciers and sea ice, particularly through satellite observation. There remains the debate about the causes of the phenomenon of global warming: it is man-made? Is it part of a natural evolution or cycle? It is both?

There is no doubt that the production of CO_2 has tremendously increased since the beginning of the industrial revolution, due to the large-scale use of fossil fuels in industry, transport, housing and leisure activities. But there is equally no doubt that the earth has gone through periods of climate change in the past, as is testified by tree rings, ice cores, fossils and geology. In those times, which date back to several thousand years, no human activity was involved in the process: a natural cause for climate change is therefore possible. It is on those grounds that Nigel Lawson, for instance, claims that it is not worth betting on a human cause and jeopardizing the western model of development which has brought about our present wealth and comfort.

These legitimate questions, which science has failed to answer unequivocally so far, are further complicated by the different prediction models that have been produced to anticipate future evolutions in global climate, their consequences, and the possible means and cost of adaptation and mitigation. Depending on the kind of data which is collected, their selection and their interpretation, different projections can be made, from evolutions that could be mastered thanks to the future progress of science to apocalyptic visions considering even the end of the human species.

In other words, regardless of the traditional suspicions of dark hidden motives lying behind the antagonistic professed positions, there is ample room for a legitimate debate about a possible link between global warming and human activities. The new phenomenon is that even the discourse of science is no longer exempt from the post-modern distrust of explanatory systems that has affected religion, politics and philosophy, as Jean-François Lyotard showed in his 1979 report published under the title *The postmodern Condition*[1]. The situation is comparable to the philosophical impasse staged by David Hume in his *Dialogues concerning Natural Religion* (1776), in which the debate degenerates into an inconclusive and unconstructive dialogue that reveals the incapacity of man's reason to establish a definitive discourse on truth.

The purpose of the present contribution is to suggest that the present predicament can be tackled under a new perspective derived from phenomenology and psychology. The aim will no longer be to reach for a positive truth about global warming, but to study the phenomena of conscience typified by the various representations of the world at play in the debate, so as to find out what they can reveal about ourselves rather than the world or the "environment". Indeed, the debate can help us to differentiate two kinds of interpretation: those that claim for certainty, and those that emphasize doubt. Historically speaking, certainty was originally the traditional characteristic of religious faith. It led to spiritual elevation, truth, crusades, the Inquisition and fanaticism. By contrast, philosophical thought began in doubt with Socrates and Plato, later on with Locke and Hume, and eventually with Heidegger, for instance. Of course, many philosophical discourses have had a claim to certainty (Descartes' philosophy of the subject, for instance, or liberal ethics, or Marxism…), but the philosophical trend initiated by David Hume, taken up by Husserl's phenomenology and, to a certain extent, by the thinkers of postmodernism, insists rather on the impossibility for man to produce a true and definitive discourse of universal explanation, so much so that some have read in Hume "the end of philosophy"[2].

In the case of global warming, we are faced, therefore, with a similar philosophical problem of doubts and certainty. This is where Carl Gustav Jung, as a psychologist, provides us with an interesting insight into the

[1] Lyotard, Jean-François: *The Postmodern Condition: a Report on Knowledge.* (1979), Manchester University Press, 1992.

[2] See : Michaud, Yves. *Hume et la fin de la philosophie,* Paris : PUF, 1983.

workings of the human mind. He started his enquiries into the human psyche as a staunch empiricist, much after the model set up by Freud, acquiring knowledge on the lockian grounds of observation (sensation) and thinking (reflection). But he soon became aware of the limits inherent in this epistemological approach and, with the assistance of phenomenology, started to consider that ultimate reality was beyond man's scope, whether it be religious or subconscious, whether it be the reality of the Self or that of the world. "We are desperately shut up in a world of psychic representations", he claimed, thus disqualifying any lockian pretence to objective perception in favour of the consideration of the conditioning role of the archetypes of the collective subconscious on any individual's perceptions, feelings, thoughts and images. Thus did he come to expose empiricism as "a general form of prejudice", just as any representation of the world or any type of "knowledge". Indeed, after an extensive study of the great discourses in the history of mankind (from the first mythologies through religious dogmas and philosophical systems to contemporary science), taking each of them as particular phenomenon of conscience with its own specific principles (which Lyotard calls "local determinism"[3]); on the ground of his analyses of his patients' dreams and artistic productions; and taking into account many works of literature, Jung was led to identify a number of significant recurring patterns which he called "archetypes", and to formulate the hypothesis of a supra-material collective subconscious, thus venturing far beyond the limits of strict empiricism.

> A general prejudice has it that the main ground for our knowledge be exclusively given from outside and that *nihil esse in intellectu quod non antea fuerit in sensu.*
>
> And yet (...) all those who know about old science or old philosophy of nature know the part of psychic data that is projected into the unknown of the external phenomenon. That part is indeed so great that in the last analysis we are never able to say how the world itself is made in the end, for we are truly compelled to transpose the physical element into a psychic process as soon as we are talking of knowledge. (...) It is the soul that provides these images and forms for us, and they are ultimately the only means of knowing the object[4].

In the light of these preliminary considerations, it seems possible, to begin with, to establish a link between positions in the global warming debate and economic motivations such as belonging to a lobby for coal or oil industry, or any industry related to these sources of energy, or being a

---

[3] Lyotard, J.F. *opus cit*, p. xxiv.

[4] Jung, Carl Gustav. *La Réalité de l'âme* Paris. Paris : Le livre de Poche, 1998.

member of a research team working in the opposite direction. Greed for money can be an end in itself that conditions one's state of mind, just as science or art:

> If science be an end in itself, man is truly right to be only an intellect. If art be an end in itself, the capacity for representation is the sole human value, and the intellect can be discarded. If greed for money be an end in itself, science and art are of no avail. No-one can deny that modern conscience is desperately torn apart by these 'ends in themselves'. Thus do men develop only specialized qualities, they themselves become instruments.[5]

The Intergovernmental Panel on Climate Change established a link between global warming and human activities as early as 1995. This link was confirmed by the American Academy of Sciences in 2001 and by a UNO report in 2002. In 2006, Nicholas Stern published his report and Al Gore's film and book *An Inconvenient truth* popularized its alarming conclusions. Each of the IPCC reports has since then confirmed that global warming is linked to emissions of green house gases by man, and warned us of the dire consequences to be expected in a not too distant future.

Despite these alarm bells, Rio, Kyoto, Bali, Copenhagen, Durban, the successive international summits that were designed to take action, have all proved as disappointing and ineffective as the French 'Grenelle de l'Environnement'. This is the problem we are facing, and we must ask why.

One first possible answer is that alarmist demonstrations have been matched by other arguments of an equally scientific claim, putting forward different readings and conclusions. In other words, science appears nowadays as incapable of producing a consensual discourse of truth, unlike what happened when heliocentrism superseded geocentrism, when evolution superseded creationism, or when the Big Bang and Big Rip theories were presented as the ultimate scientific accounts of the universe and its evolution through time and space. Of course, science is periodically subject to revisions and innovations: from Euclidian geometry to relativity and quantum mechanics with its universe of *n* dimensions, for instance. Yet, what is at stake in the debate about global climate has nothing to do with a conceptual revolution. The heart of the matter is observational data and their interpretation (ice cores, tree rings, the melting of glaciers and sea ice, temperature average values, etc). And the paradox is that it is when its tools of observation have become most performing (think of satellites, for instance) that science turns out to be the least able to provide

---

[5] Jung, *opus cit.*, 972-3.

a unifying representation and consensual understanding of the observed phenomena![6]

It appears that the disturbing element here is the eminently constructive character of the scientific outlook, or rather of the different scientific outlooks. Moreover, we are not dealing only with science, but also with economics, politics, ethics, and psychology: fields of knowledge about which Jung made the following statement in his Tavistock lectures:

> The powerful factor that changes the whole life, transforms the face of the earth and makes history is the collective psyche. And the collective psyche follows laws that are totally different from those of our conscience: the great decisive forces are the archetypes, they are the ones that beget the events of the real world, not our reason nor our intellect.[7]

Hence the following invitation to tackle the global warming debate not by comparing the different scientific arguments but, firstly, by assessing their respective grounds and use of rhetorical tools, secondly, by studying the implied modes of conceptualization and archetypal manifestations in the various discourses, in order, thirdly, to eventually insist on the necessity for a change of perceptive and mental paradigms.

The criticism of grounds

Al Gore's movie, which was released in 2006, popularized the theme of global warming. One of its most striking features is the deliberate blending of scientific arguments and rhetorical devices that aim at rousing fear in the public or seducing it.

In the first category we find for instance the parallel curves of CO_2 emissions and world temperatures, statistical data concerning CO_2 emissions going back 600,000 years to show that the levels have never been above 300ppm in the past, unlike today; the observed changes in migration and nesting patterns for birds and in the cyclical return of some insects such as caterpillars and ladybirds; changes in the geographical distribution of mosquitoes; and the rise in insect-born infectious diseases. These charts, though addressed to reason, are sometimes presented in a

[6] By comparison, one should bear in mind that the widely accepted big bang and big rip theories are based on the observation of only 4% of the universe, if dark matter and dark energy are not only the necessary concepts for our new dynamic vision of the universe, but do indeed exist and account for 96% of the universe!)

[7] Jung, C.G. *Sur les Fondements de la psychologie analytique : les conférences de Tavistock*. Paris : Albin Michel 2011, p. 230-1.

dramatic fashion, as when Al Gore presents the famous hockey stick curve of CO_2 emissions (from the year 100 to the year 2000) standing on a lift truck. The aim obviously shifts from rational understanding to emotional response.

The argument is from now on accompanied with impressive pictures. Melting glaciers falling into the sea, hurricanes, typhoons and floods (in India and China, for instance) soon take on apocalyptic dimensions, with alternate photographs of desertification in the Nile valley, lake Tchad and the Aral sea. Widening cracks in the sea-ice of Arctic and Antarctica, so-called "drunken trees" and collapsed buildings as consequences of the softening of permafrost echo the dramatic visions of deforestation in Haiti and Brazil. To complete these worrying documents, a cartoon vividly pictures the foreseeable floods that will submerge Florida, San Francisco, Bejing, Shangai, Calcutta, Bengladesh and, last but not least, Manhattan, where even the memorial to the 9/11 tragedy would not be spared by a six meter rise in the sea level. In other words, symptoms (such as the melting of glaciers and sea ice), consequences (such as extreme weather events) and causes (such as deforestation) are presented in a chaotic way leading from documentary evidence to fiction. In keeping with this intent to stir up emotions rather than cause reflection, these passages are interlarded with deeply personal sequences during which the lecturer introduces us to the farm in which he lived as a child, his feelings of unfairness when he was defeated by Bush in the 2000 presidential election, his son's narrow escape from an accident, and the loss of his sister to lung cancer, which he felt all the more deeply as his family were tobacco producers. These moments, which have little to do with global warming, are meant to suggest depth of thought in face of life's trials, a sense of responsibility and a certain personal ethics in the speaker. They probably contribute to giving more weight to his conclusion on a real consensus on global warming in the scientific community: there is "virtually no disagreement", he claims, since out of the 928 articles that had been published on this subject in 2006, none seemed to deny the phenomenon. At this stage, the fiction that completed the documentary evidence leads to wishful thinking, as the subsequent public debate about global warming, its causes and consequences, was to reveal.

A great number of publications have indeed, since then, either denied the evidence about the rise of global temperatures, or the link with human activities. For instance, in France, Claude Allègre published *Ma Vérité sur la planète* in 2007, and in the United Kingdom, Nigel Lawson wrote a book significantly entitled *An Appeal to reason* in 2008. A number of articles and television programs, such as *The Great Global Warming*

Swindle, which was broadcasted on Channel 4 in March 2007, also came to contradict both Al Gore's movie and the Stern report. The debate became fierce without gaining in clarity. Indeed, it is the mark of an absence of critical spirit that some of the participants who were no specialists of global warming could enjoy a publicity in the media at least equal to the that given to the reports of the IPCC, which are written, checked and approved by an international community of scientists who are experts in their field of study related to climate change: part of the problem is that authoritative experts are not heard[8].

It is also worth noting that the authors who deny the reality of climate change or its relation to human activities use as many rhetorical devices as their adversaries, and also aim at feelings rather than reason. The supposed threat of degrowth is presented as a return to candlelit times, the cost of mitigation through the reduction of CO_2 emissions and the shift to alternative energies is said to be so high that it will lead to frightening economic consequences (this was before the 2008 financial crisis!), and it is suggested as a solution that we have faith in the future progress of science to solve the problem of green house gases with the development of carbon capture and storage techniques, the diffusion of other gases in the atmosphere to counterbalance for the green house effect, or the proposed positioning in space of immense screens designed to reduce the amount of solar energy reaching the surface of the earth! The discourse insensibly passes on from fiction to science fiction.

The lack of scientific grounds in some arguments has sometimes led to the discovery that their authors belonged to oil and coal industry lobbies, as was the case, for instance, of Philip Cooney, whose career as government official in charge of environmental questions under Bush and then as executive in Exxon Mobil, is exposed by Al Gore. A discourse of "science" can sometimes hide non-scientific motivations: but then, who is truly in a position to establish what the scientific truth is about global warming? The main point, for the likes of Philip Cooney, is simply to cause enough doubt for the general public to remain indifferent to what appears in the eyes of committed ecologists as a shocking lack of political will and measures to counteract the catastrophic consequences of man-induced climate change.

In this respect, things have only become worse since the 2008 financial crisis: today, in 2012, the media are more concerned with the euro crisis, world finance, the rise of unemployment, matters of living standards,

---

[8] Allègre, for instance, is a geologist; Lawson is a former Chancellor of the Exchequer; and even a popular TV weatherman felt he had to publish a book on the topic: Laurent Cabrol.

education, health, housing, and so forth, than with global warming. But beyond this observation, it becomes clear that we are indeed dealing with phenomena of conscience in the phenomenological sense, hence the paradox of asserting strong positions on both sides, to the point of rigid confrontation, in a context in which our only possible objective element of certainty is this impossibility to establish a universal consensual scientific truth!

Our observational grounds are limited: 3000 Argos buoys are supposed to monitor 360 million km^2 of seas and oceans, that is, one for each 100,000 km^2: would five temperature readings give us an significant average of the temperature of an area as wide as the French territory? Has the figure for an average sea level any meaning when we consider that the mass of water that covers 70% of the earth is never stable but endlessly moving because of waves, tides, irregularities in the earth's gravitational field, etc? What degree of precision can be ascribed to sea level readings made in the past centuries? Were temperature readings any better, even in under-developed countries? Given that the earth's atmosphere is even more unstable than the sea, how can we assert that the world average temperature has risen by 0.2°C since 1998? What is the degree of precision of temperatures derived from the abundance of oxygen isotopes in polar ice cores, or from the thickness of tree rings? Can the average rate of CO_2 in the world's atmosphere really be measured with a precision of a few parts per million? Were extreme weather events recorded with the same accuracy as today in the past centuries, worldwide? And when we observe an increase of such events over the past ten years (with an acceptable degree of accuracy), who is to say whether it betrays a long-term trend or is simply a statistical accident in a period of several thousands of years? When Al Gore observes that insurance claims have gone up because of the increase in violent weather events, does he take into account the fact that insurance cover is much more widespread today than in the past? What confidence can we place in figures that are based on such moving grounds? What is the true representative value of these average figures and curves imposed on a characteristically fluctuating reality?

Modes of conceptualization

The problem, in the end, the space in which doubt can grow and flourish, consists in that these observational data and their interpretation are widely beyond the scope and reach of the general public. It is the reserved field of scientists and experts. We need to trust scientific experts:

the problem is that they disagree. Take for instance the parallel curves of CO_2 density and atmospheric temperature. As far back as we can go in the past (that is, roughly 600,000 years by Stern's assumptions), these two curves follow similar cyclical patterns of rise and fall. For some scientists, the evolution of temperatures follows that of CO_2 concentration in the atmosphere, thus suggesting a causation relationship, but to others, it is the other way round: CO_2 density increases or falls down *after* temperatures have done so: in that case, it is the temperature rise that causes the CO_2 rise!

In face of such contradictions, and in the absence of objective undisputable data and conclusions, the average layman (if there exists such an individual) can only rely on a personal act of faith. David Hume showed in the 18th century how all our beliefs are ultimately grounded in a mere feeling of truth, and how a belief can be defined as an "act of the mind that gives more weight to realities than fictions"[9]. Thus, a mystic will give more weight to spiritual realities than mundane illusions, while a materialist will set a higher value on objects of sense perception, sometimes going so far as to deny the existence of any supra-material "reality"[10]. This psychological process accounts for the use of rhetorical devices which we identified in Al Gore's movie[11], and anticipates C.G. Jung's analysis of the workings of man' conscience when enclosed in world of psychic representations: the more deficient and doubtful the empirical data, the greater the part of speculation through which archetypes predetermine thoughts.

Accordingly, taking into account the way an individual subject constructs his own "object" of observation through his constituting

---

[9] See Hume, David: *A Treatise of Human Nature* (1740) Oxford, Clarendon Press, 1983, pp. 103 and 629.

[10] Thus, for instance, William Blake's conviction that « mental things are alone real » stands in direct contradiction with scientist Jacques Monod's postulate of the objectivity of nature…

[11] In a parallel way, John Locke, when he wrote *An Essay concerning Human Understanding* to lay the foundations of empiricism and the modern scientific approach, was confronted with a well-established tradition of philosophy based on innate ideas. The problem was (and still is) that no empirical proof can be given of the non-existence of innate notions in man (which made it possible for Jung, after growing out of empiricism, to resort to the ancient notion of an innate legacy in man's psyche). Locke, when faced with the impossibility of contradicting the tradition of innate ideas through philosophical argumentation, resorted to rhetorics in order to impose his new empirical perspective: see my paper « La Sensation et la réflexion selon John Locke, ou la rhétorique au service de la séduction » *XVII-XVIII*, n°65 (2008), 131-153.

outlook, we can regard this "object" as a mirror image of the subject[12]. The traditional division between the observing subject and the observed object is superseded by an inner split within the subject as he takes him/herself as an "object" of self-knowledge. In Jung's words,

> We can fully behold ourselves only the mirror of the image of the world that we build for ourselves. We appear only in the image which we create. It is through our creating act that we come into full light and can know ourselves as a reality.[13]

A number of archetypes are most conspicuous in Al Gore's movie, particularly when he explicitly refers to biblical themes. For instance, one of his comments about the apocalyptic images we have listed is: "There will be a day of reckoning". Conjuring up the image of the Last Judgment implicitly calls for that of an original sin and relates to what Thomas Molnar termed our "guilt of Being", an archetypal feeling of guilt[14]. The planet earth is presented through two significant pictures: as a blue sphere when seen from the moon, and as a minute dot lost in space captured by the Voyager probe. They convey the ideas of the One, the Whole and nothingness, those inter-related archetypes in the collective subconscious which are commonly summoned by mystics when referring to the divine.

Thus does the planet soon assume the status of a deity. Behind Lovelock's Gaia stands the archetype of the Great Mother whose manifestation produces the antithetical figures of the protective Mother and the devouring Mother. The latter is the result of the archetypal original separation and subsequent guilt (think of the closed gates of paradise), while the former was the nourishing Garden of Eden, whose trace lives on in the pastoral genre. This basic duality is to be related, according to Jung, to the double-faced god of the Gnostics, and, more fundamentally, to man's own "paradoxical nature". With the devouring mother, evil is no longer exclusively man-made. Natural disasters (such as the 1755 Lisbon earthquake, which Voltaire opposed to Leibniz' optimistic philosophy) are

---

[12] The notion of « constituting outlook » is developped by Guy Lardreau in *La Véracité* (Paris, éditions Verdier, 1993) on the grounds of Hume's feeling of truth.

[13] Jung, *La Réalité de l'âme*, p. 975.

[14] See Molnar, Thomas: *Archetypes of Thought* (London: Transaction Publishers, 1996), pp. 27-48, in which Molnar traces this archetype from Parmenides to modern philosophers, through Judaism, Christianity and Gnosticism: "Whatever claims to exist (…) [has] exited from Being. The multiple, the world of motley phenomena, issues forth from a fall from compact Being into diverse individualities. The latter are subjected to punishment for their exit and they expiate their guilt of breaking away by immersion in *time*." (p. 29)

part and parcel of the natural course of events, through which nature is no longer a cocoon to man but the source for a feeling of otherness and strangeness. A radical split thus occurs between the subject (man) and the object (nature), and the gap will only widen with the rise of science and technology.

Francis Bacon, in the early 17th century, stated that the experimental method would help man "to restore dominion over nature". Indeed, the industrial revolution of the 18th century has as it were made up for the loss of Paradise by enabling man to domesticate nature and exploit its resources. The western notion of "progress" appears as a revenge on the consequences of original sin: this is the subconscious archetypal root of the economic development that has led, in the 20th century, to the consumer society in the wake of the Glorious Thirties, and the global expansion of liberalism known as "globalization".

The new "utilitarian" relationship to instrumental nature calls for another archetype: that of Prometheus. Man now controls and exploits nature through irrigation techniques, dams, and mines. He uses coal, gas and oil as sources of energy to transform raw materials into useful artifacts that can be processed, transported and sold on the market. He has even domesticated divine fire in his nuclear plants, thanks to atomic fission, and now aims at creating solar energy in his laboratories through atomic fusion. Thus capturing divine fire for his own benefit (whether from coal, oil or atom), he follows the example of Prometheus, who stole the divine fire to give it to man.

But Prometheus was severely punished for his deed: he ended up chained on a rock for an eagle to feed daily on his liver. Whether, as some contend, he anticipated the sacrifice of Christ or not, he undoubtedly refers to an archetypal feeling of guilt. This may be why the Prometheus figure of the rebel soon decays into that of the sorcerer's apprentice. It is now becoming clear that the expansion of the western model of development inherited from the Enlightenment and the industrial revolution is a considerable threat to our environment. Phenomena of pollution and acid rain, the hole in the ozone layer, the depletion of mineral and energy resources, and, last but not least, climate change account for Al Gore's dramatic statement: "We are on a collision course between our civilization and the earth". The feeling is no longer simply one of strangeness, but of total opposition. Nature is angry at us, as his movie shows through pictures of tsunamis, flooded polders, destructive cyclones and typhoons. It has now become a hackneyed theme in the ecological movement that it is impossible to contemplate an infinite progress of the production-distribution-consumption spiral within the context of a limited planet with limited resources. Two more planets would be necessary if the western model of development and standards of living were to be adopted by so-called "emerging nations"!

Man's conquest of nature has thus turned into total alienation. Far from being a beneficent and nourishing matrix, nature has become an "environment", that which simply surrounds the center that man implicitly claims to be. By aspiring to reach beyond his natural condition, through his mastery of nature, man posits himself as the center and as an end in itself. This symptom can be related to what Jung identified as man's loss of his subconscious roots. In the traditional mythological visions, man was part of nature, gods were present in nature, and man knew the meaning of his life within this harmonious whole. By contrast, modern astronomy, as the latest achievement of the scientific *logos*, is unable to convey any meaning. Not only has it expelled man from the center of creation with the heliocentric vision, but it also tells the story of the universe from its beginning (the Big Bang) to its end (the Big Rip), in an account devoid of any teleological sense: it is fundamentally tragic and absurd, since it ends in complete annihilation[15].

The ensuing neurosis may account for the denial of reality which characterizes those who profess "business as usual", in the name of an irrational faith in the capacity of science to find solutions to global warming. This position, which is exemplified by Nigel Lawson, is based on the repression of the archetypal feeling of guilt. A process of identification with the Prometheus archetype leads man to have faith in the future achievements of science, but at the same time leaves man in a meaningless world. By contrast, one could interpret the idea of degrowth advocated by Serge Latouche as a form of return of the repressed, and a case of identification with the *sentenced* Prometheus, called to humbly abandon his ambitions[16]. This is where Prometheus is linked to the

---

[15] See Steven Weinberg :

Il est quasiment impossible aux êtres humains de ne pas croire qu'il existe une relation particulière entre eux et l'univers, que la vie n'est pas seulement une suite grotesque d'accidents remontant dans le passé jusqu'aux trois premières minutes, mais que, d'une certaine façon, nous fûmes conçus dès le commencement… Il est plus difficile encore de réaliser que cet univers a évolué à partir de conditions initiales si peu familières qu'on peut à peine les imaginer, et doit finir par s'éteindre dans un froid interminable ou dans une chaleur d'enfer. Plus l'univers nous semble compréhensible, et plus il semble absurde.

quoted in Nicolescu, Basarab *: La science, le sens et l'évolution. Essai sur Jakob Boehme (Paris, Félin, 198), p. 55.*

[16] see Latouche, Serge : *Petit Traité de la décroissance sereine,* Paris : Mille et Une Nuits, 2007.

archetype of the Mother. Knowledge and mastery of nature are divine prerogatives that lie beyond the scope of mankind because they are the prerogatives of a jealous god.

The two main parties in the debate about global warming can therefore, in the last analysis, be referred to archetypal guilt and desire for the mother. Now, Jung read the idea of non-duality in the Advaita Vedanta as a case of "nostalgia for the lost Mother" made manifest in "the longing of the unconscious for the light of consciousness" met by "the aspiration of conscience for the deep well of being and the communion with life under all its innumerable forms"[17]: from this point of view, the hindu fusional vision is no longer *experienced* but *conceptualised* as the translation for a desire for the Mother. Its fulfillment implies the transgression of the incest taboo, leading to a second birth, the birth of the Self or the future personality. Master Eckhart provides a vivid example of this process:

> When the soul reaches the One and enters there through a clear rejection of itself, then does it find nothingness, as it were, in God. It appeared to a man, as in a dream – it was a waking dream – that he was expecting nothingness as a woman expects a child, and in this nothingness God was born; he was the fruit of nothingness. God was born in nothingness[18].

Archetypes gather and act in constellations: the loss of the protective Mother-nature gives rise to a feeling of guilt that transforms the Mother into a devouring figure. Divine prerogatives are now forbidden to man: the fruits of the tree of knowledge are forbidden, and when man attempts to steal them, the gates of paradise close on the lost Mother and the incest taboo prevails. Subsequent efforts to acquire divine powers are severely punished by jealous Gods, hence the tragic fate of rebellious or ambitious figures such as Prometheus, Icarus, Acteon, Semele, Tentalus, Pheaton, and Bellrophon, among others. In the same way, in the Bible, acts of disobedience on the part of Israel are punished by the God of the Old Testament, whose status as a god of anger entails the tropes of the Last Judgment and the Apocalypse. Al Gore thus appears in his movie like a modern-day prophet roming the world to spread the word to a people that does not heed the warning: "There will be a day of reckoning!" The very title *An Inconvenient Truth* tells of the psychological stake of his mission: a questioning of our habits of thought, a transformation of the psyche similar to Job's experience of God in the whirlwind. Al Gore addresses those who idolize science and profit without any consideration of nature or

[17]Segal, Robert, A. ed. *Jung on Mythology*, London: Routledge, 1998. 148

[18] Maître Eckhart, *Le Détachement*, (XIVe) Paris : Payot, 1995. 97

the future generations. When antagonism has superseded harmony, the separation from the Mother is completed, and the devouring figure calls for the archetype of the apocalypse as revelation and transforming experience. This is how the poet William Blake read the Last Judgment, this is why David Hume calls for further revelation at the end of his philosophical dialogues, thus anticipating Hubert Reeves's comparing man's knowledge, in front of "the wall of ignorance" which bars access to "what was before the Big Bang", with that of the first prehistoric hunter[19]. This is the modern version of the forbidden fruit of knowledge and closed gates of Paradise.

On this stage, a new play is being performed, with new characters, deriving from another type of conceptualization. Within this perspective, the prime objective of progress as it is understood in the western world is to establish man's domination over nature. Through the promethean goal of the development and science and technology, this process leads to an overexploitation of nature which can be read as a manifest will to power[20]. Contrary to an ethics of humility and resignation, defeat of the ego and non-action, the will to power is the ground for faith in science and technology, from their developments prior to and through the industrial revolution, to their future solutions in CCS (carbon capture and storage), together with the diffusion of gases in the atmosphere or the setting of giant screens in space to counterbalance the consequences of the rise of CO_2 concentration.

In thus passing from fiction to science-fiction, advocates of the will to power often lose sight of ethical considerations. Nigel Lawson, the former Chancellor of the Exchequer under Margaret Thatcher, thus puts forward the idea of a rational management of risk by virtue of which it would be suicidal to change anything in the course of our present economic model as long as the link between global warming and human activities has not been established beyond doubt. In this frame of mind, ethical arguments about applying the precautionary principle for the sake of the wellbeing of poorer countries (those that will be first affected by climate change) or future generations will have little bearing.

---

[19] Hume, David. *Dialogues concerning Natural Religion* (1777). London : Penguin, 1990.

Reeves Hubert. *Patience dans l'azur.* (1981) Paris, Editions du Seuil, 1988 : p. 68 : Le seul vrai 'problème', c'est celui de l'existence même de l'univers. 'Pourquoi y a-t-il quelque chose plutôt que rien ?' Sur le plan scientifique, nous sommes incapables d'y répondre. Après plusieurs millénaires, nous en sommes ici au même point que le premier chasseur préhistorique venu : au zéro absolu.

[20] See Nietzsche, *The Will to Power*, (1930) New York, Random House, 1968.

But most of all, it is by contrast with eastern philosophies that the main features of western thought become salient. In the Hindu tradition, for instance, the primal experience of being, the experience of the One, the whole, and nothingness, implies a totally different perception of the relationship between man and nature. The economy in point is neither liberal nor materialistic; it encompasses the supra-material dimensions that the West has lost. No ethics of action can divert the spectator's conscience from the experience of the world and others as manifestations of being and transcendence. Man's conscience mirrors this manifestation, it is never an end in itself. Thus did Heidegger find in Oriental studies the opportunity for a radical rethinking of the western mode of conceptualization that conditions our relation to nature. He traces back its beginnings to the pre-Socratic thinkers and shows how it was mainly with the Roman reading of Plato's and Aristotle's works that a conceptual frame was built on the basic opposition between subject and object, and on the law of causation as a structuring principle. In Heraclitus and Parmenides, Heidegger rediscovers the initial experience of being, which enables Brad Bannon to establish strong thematic parallels between Heidegger's ideas and the *Advaita Vedanta* tradition[21]. Indeed, the German philosopher regards the ruin of the planet as the consequence of man's will to power, and absence of meaning as the result of the conception of the earth as "a wandering body", which echoes William Blake's exposition of the astronomical vision of the earth as "a globe rolling through voidness"[22].

> Only the will, with its firm footing in technique, shakes the earth and causes it to wear away in fatigue and the variations of the artificial. It forces the earth to reach beyond the limits of the sphere of possibility that has developed around itself and pushes it into what is no longer the possible, and which, therefore, is the impossible.[23]

It is by reference to the same will that some have faith in so called "sustainable development". In this respects, Heidegger's warnings sound prophetic: "That technology produces many inventions and a constant string of new artefacts is no proof that its conquests may change the impossible into the possible."[24] Instead of an irrational faith in the future

---

[21] See Bannon, Brad, *The Quest for Postmodern Ethics, a Phenomenological Comparison of Heidegger and Aurobindo.* Bangalore: Dharmaram Publications, 2007.

[22] Blake William, *Milton,* book 1, plate 29, l. 16

[23] Heidegger, Martin, *Essais et Conférences,* (1958, Gallimard, 1997), p. 113. See also Heidegger's *Heracliteus* and *Parmenides.*

[24] Ibidem, 113-4

capacity of science to make up for the negative consequences of technological progress, what is needed is a change of mental paradigm.

Change of perceptive and mental paradigm

In Heidegger's words,

> It is one thing to simply take advantage of the earth. It is another to receive its blessing and, little by little, feel at home within this conception law, so as to keep watch over the secret of being and preserve the sacredness of the possible.[25]

It is clear that the modern version of the will to power, by contrast, and despite Nietzsche, posits man "*hic et nunc*" as an end in himself, which, given his mortal condition, is necessarily meaningless, and turns out to be destructive. Such is the outcome of an over-confident science:

> The devastation of the earth begins with a deliberate process which is not and cannot be known in its being. It begins at a time when the being of truth is defined as a certainty, as that through which human representations and productions become self-confident. Hegel conceives this moment in the history of metaphysics as when absolute self-conscience becomes a principle of thought.[26]

One could read the Gaia cult that defines the earth as a deity as a form of compensation for such a unilateral conscience, and degrowth as an emphasis on limits based on acknowledgement of the archetype of death. In this perspective, the transforming experience does not aim simply at the relationship between man and his environment, but involves first and foremost the inner depths of the subject. It is because of his awareness of death that man cannot be content with simply following his chthonic bend. He also develops a spiritual yearning for a victory over death: this is the spiritual side of the paradox of human nature which Jung identified as a basic split in man's nature.

Ethics of action based on the will to power belong to the chthonic side, and activism or entertainment in Pascal's sense can help to ignore the fear of death. Myths, religions and the archetype of resurrection talk to the spiritual side: their comfort alleviates the fear of death. Yet, both psychological states are rooted in that basic fear, which characterizes

---

[25] Ibidem, 114
[26] Ibidem, 115

human conscience. Even scientific predictions about the end of the earth and the sun in five billion years, or the end of the universe in infinite dilution, even models predicting climate catastrophes that could jeopardize the human species itself can be seen as cases of projections due to the return of the repressed fear of death. In the same way, anticipations of an apocalypse of the nuclear kind during the cold war, or according to the Maya calendar for 2012, are indicative of the present-day loss of efficiency of traditional symbols of immortality, soul and resurrection.

Jung's analysis of the civilization *malaise* of the West runs along the same lines: he identifies the roots of our collective neurosis in the split between a modern conscience under the spell of scientific and technological progress, on the one hand, and, on the other, the deep roots of conscience in the supra-material collective subconscious. It is indeed remarkable that the development of empiricism by Locke on the grounds of Galileo's and Newton's scientific discoveries began with an *a-priori* discarding of the innate and the supra-material. It is therefore legitimate to read the inconclusive debate on global warming not only as a "crisis of European sciences" (to take up Husserl's phrase[27]), but also as the symptom of a crisis of the psyche in the western world. This crisis can be felt through the evolution of the philosophical quest for truth: Heidegger feels the need to go back to pre-Socratic thinkers because he sees in the subsequent modes of conceptualization the mark of the loss of the initial experience of being. With Socrates, the form of the dialogue provided the means for a philosophical quest grounded in innate ideas, but with David Hume, in the 18th century, the dialogues become a synonym for a philosophical dead end which anticipates the postmodern crisis and its definite impossibility to put forward a universal discourse on truth.

In the 20th century, nevertheless, Jung advocates a new form of dialogue, an inner dialogue with inner voices and characters. With *The Seven Sermons to the Dead* or the recently published *Red Book*, he provides numerous examples of deep inner dialogues and their role in a psychological quest which is also philosophical and theological[28]. His hope is that each individual will open himself up to a new transforming experience: the meeting with the subconscious. Following Paul's conversion on the way to Damascus, the moment of illumination of the mystics and the apex of Hindu contemplation, such an experience begets a

[27] See: Husserl Edmund. *The Crisis of European Sciences and Transcendental Phenomenology* (1954) Northwestern University Press, 1970.
[28] Jung, C.G. "The Seven Sermons to the Dead", in *La Vie Symbolique* (Paris: Albin Michel, 1989), and *The Red Book,* (1957) New York: W.W. Norton & Company, 2009.

new type of relationship to the world, to others, to transcendence, and to meaning. The problem, from the point of view of conscience, is that it lies beyond thought and language, in a dimension in which the One, the Whole, Being and Nothingness are no longer irreconcilable conceptual categories. We are confronted with the shortcomings and limits of our understanding and language. In other words, this type of inner experience spells an enlargement of conscience, as is testified by the accounts of mystical literature or eastern philosophies. Thanks to these texts, even if the experience *per se* cannot be reached through intellectual means or academic discourses, it can nevertheless provide a point of reference by virtue of which a number of questions can be raised in order to address the supposed certainties of science.

Immediately after the systematization of the empirical method by Locke, David Hume developed a thorough criticism of the limits of the new kind of knowledge, and, in his *Dialogues concerning Natural Religion* observed the incapacity of human reason to produce a discourse on meaning that would be universally acknowledged. He was thus led to shift the emphasis onto the roots of the multiplicity of discourses, namely the subjective feeling of truth that lies at the bottom of any belief and subsequent acts of the mind. This analysis was taken up by Husserl in *Crisis of European Sciences*, and, more recently by Guy Lardreau in his comparison between the philosopher's *experimentum mentis* and the mystic's illumination[29]. Yet, it is by ignoring these reflections that science and technology became the grounds for the western model of development and its notion of "progress". This model today has to face its limits, not only because of the global warming crisis looming large on the horizon, but also because the planet cannot provide the resources that would be necessary for the expansion of this model to all nations. The inconvenient truth leads to an inconvenient question: what meaning can the west stand for?

The psychological crisis of the west falls under three headings at least. The first one is the environmental issue: a model based on a continuous growth of production and consumption is simply not sustainable on a limited planet. The second heading is the financial predicament of individuals, nations and groups of nations being led to spend more than they earn, and, when resorting to credit, more than they can possibly pay back, a situation the roots of which had been exposed by Marx in his

[29] Lardreau, Guy, *Discours philosophique et discours spirituel : autour de la philosophie spirituelle de Philoxène de Mabbourg*, Paris, Seuil, 1985. p. 130.

analysis of the inherent contradictions of the capitalist system[30]. Note that within this economic field, global warming is labeled as "a market failure", and consequently cannot be dealt with through the market forces. Only the state can take charge of this shortcoming[31]. Hence a paradox: as claimed by its name, liberal economics rely on individual freedom, and freedom of enterprise is usually linked with the idea of democracy. The modern state should therefore be an expression of the general will of the people, just as the market should translate the free interaction of offer and demand. Yet, when the state, in the name of environmental concerns, imposes restrictive legislation on economics, it seems to go against the liberal principle of freedom. This case of collective schizophrenia is based on the idea that the state would be rational, whereas the market forces and international finance would be the result of irrational motives and behaviour. From a psychological point of view, one could easily identify these grounds with what Jung terms the shadow, as Michel Maffesoli's works amply demonstrate[32].

In the same way, the process of deindustrialization in the West, which transfers much of the production to countries in which the socio-political conditions make it possible to derive bigger profits from the exploitation of the working class, is further confirmation of the fact that moral principles cannot stand in front of economic and financial considerations. This also accounts for the fact that despite the supposed and much publicized reduction of poverty world-wide, the gap between the wealthy minority which benefits from the growth of so-called emerging countries and the majority of the poor has never been so wide, whether it be inside those countries or between the nations of the world. In emerging countries, the majority of the population does not benefit from the process of intensive industrialization, while in the post-industrial West, the redeployment of activities towards the production of luxurious goods aims at cornering the market of the new wealthy minority in the BRIC countries: the circle is complete and demonstrates that this economic reorganization of the world feeds on worsening inequalities, regardless of

---

[30] The credit system of the subprimes is a direct consequence of the contradiction between the interest of the capitalist in paying as low wages as possible to his workers, and the necessity to keep a market with sufficient demand for the produced goods to be sold with a profit. Subprimes made it possible to keep wages at a relatively low level while artificially developing the purchasing power of consumers… until the bubble burst.

[31] See Von Weizsäcker, Ernst Ulrich. *Factor Five,* London: Earthscan, 2009

[32] See for instance : Maffesoli, Michel. *La Part du Diable*, Paris, Flammarion, 2002.

any moral or humane considerations. How then could we expect this system to pay any attention to the fate of future generations?

These crucial issues of the post-modern world should be at the heart of the concerns and missions of universities. In the West, they are all too often made subservient to practical training policies which ignore the financial, ethical and environmental failures of the predominant process of globalization. Fortunately, some university research centers, particularly in India, have begun a thorough critical analysis of the ethics of globalization[33].

The present study has attempted to identify the subconscious psychic elements that lie at the root of the present-day debate on global warming. Beyond economic and ideological motivations, they provide the archetypal conditioning of our relationship to nature. It soon became apparent that the very notion of archetypes entails a necessary reflection on our modes of conceptualization, which Jung and Heidegger, among others, have developed. To the latter, concepts tend to freeze what was, to pre-socratic philosophers such as Heracliteus, only permanent change and flux, while in the eyes of the former, the numinous experience of the archetype cannot be apprehended by conceptual or discursive thought. In this light, the present-day claim to set an average value on such fluctuating realities as sea-level or atmosphere temperature seem as preposterous as any attempt at putting mystical experience into words. Taking account of the limits inherent to our mode of conceptualization leads to questions which are indeed "inconvenient" (like Al Gore's truth), in so far as it sets the environmental issue in the larger context of a psychic crisis looming behind the financial, economic, social and ethical problems of our postmodern world. What is needed is a transforming experience similar to Heidegger's experience of being, or the experience of non-duality of the Advaita Vedanta in eastern philosophy[34].

To Jung, it is an experience which depends on the faculty of intuition, the channel for the nourishing energies coming from the subconscious. Only such an experience, which Jung associates with the defeat of the ego, can provide a new outlook on life, a new paradigm of thought, a new type of relationship to nature[35]. In the case of psychology as in that of

---

[33] See for instance : Parapally, Jacob and Antony Kalliath, ed: *Theology of Economics in a Globalized World*, Bangalore, Indian Theological Association, 2010.

[34] See Brad Bannon, *opus cit.* and Kachappilly, Kurian, *Process, Implications and Applications,* Bangalore, Dharmaram Publications, 2006.

Heidegger, it is an experience which can restore the feeling of harmony with the world, with others, and with the transcendence that provides meaning. The philosopher's words deserve meditation:

> No change can take place without an escort that first shows the way. But how could it draw near, if the Advent (*Ereignis*) does not begin to shine, which, by calling man's being and providing it with presence and protection, apprehends it in sight, that is in an outlook, and through this outlook leads some mortals on the way to poetic and thoughtful habitation? (...) To apprehend in sight (*er-blicken*) here means to penetrate through sight that which, in the seen thing, turns its outlook (*anblickt*) towards us...[36]

But beyond this, and beyond the innocence of yet another archetypal idea of pastoral harmony with nature, such a moment can be the ground for a critical outlook on man's over-exploitation of the planet and its resources, the growing inequalities inherent in global liberalism, the nonsensical aim of a never ending economic growth. In this respect, it can ultimately work towards the integration of man's shadow into conscience. Whether it be identified as an archetype manifested through the two-faced god of Gnosticism or as a projection of the paradox of the human soul, it is by acknowledging the shadow that man can hope to bound and bind it, rather than run the risk of its return under the violent forms of repressed energies.

How can such an experience be the subject of education? Such is the challenge which must be faced today by so-called "humanities", at a time when the incapacity of science as a source of certainty is amply demonstrated by the doubts that undermine the global warming debate.

[36] Heidegger, *Essais et Conférences,* (1958) Paris: Gallimard, 1997, p. 115.

CONTRIBUTORS

Frédéric ARMAO is a Senior Lecturer in British and Irish studies at the University of Toulon, France.

Camille BIROS is a Senior Lecturer in English for Science at the University of Grenoble 1, France.

Stephanie BORY is an Assistant Professor in British Civilisation in Jean Moulin-Lyon 3 University, France.

Neil CARTER is Professor of Politics at the University of York, United Kingdom. His main research has focused on environmental politics and policy, and on British party politics. He is currently Chair of Departmental Research Committee.

Muriel CASSEL-PICCOT is a Senior Lecturer in British contemporary history in Jean Moulin-Lyon 3 University.

Olivier GOUIRAND holds a PhD in English Linguistics and is a Senior Lecturer at Toulon University, France. His current research is centered around the semantics of environmental discourse.

Hélène LEDOUBLE is a Senior Lecturer in English at the University of Toulon and works in the field of Applied Linguistics related to the Environment.

Gilles LEYDIER is a Professor of British Studies at the University of Toulon. He is the editor of the scientific review *L'Observatoire de la société britannique.*

Alexia MARTIN is a Senior Lecturer in British and Irish studies at the University of Toulon.

Frank McDONALD is Environment Editor of *The Irish Times* and author of several books, including *Chaos at the Crossroads* (with James Nix).

Patrick MENNETEAU is a Professor of Cultural Studies and XVIIIth century British literature in the University of Toulon. He has also taught in the universities of Aix-en-Provence, Mulhouse, Grenoble, Glasgow and Aberdeen.

Pierre-François PEIRANO teaches at the Department of English at the University of Toulon. His research fields include the history of ideas, as well as the study of the representation process.

Coralie RAFFENNE is a Senior Lecturer at the University of Orléans (France) and a member of the DEMOENV research project: "Democracy facing environmental stakes".

Clare SIBLEY-ESPOSITO is a member of the Babel research laboratory in the University of Toulon, where she teaches English language and literature.

Fiona SIMPKINS is a Senior Lecturer in contemporary British History at the University Lumière of Lyon, France.

Nicholas SOWELS is a Senior Lecturer in English for economics and the study of the "Anglo-Saxon" economies, at the University of Paris 1 Panthéon-Sorbonne, France.